COMMUNITY ENGAGEMENT IN POST-DISASTER RECOVERY

Community Engagement in Post-Disaster Recovery reflects a wide array of practical experiences in working with disaster-affected communities internationally. It demonstrates that widely held assumptions about the benefits of community consultation and engagement in disaster recovery work need to be examined more critically because poorly conceived and hastily implemented community engagement strategies have sometimes exacerbated divisions within affected communities and/or resulted in ineffective use of aid funding. It is equally demonstrated that well-crafted, creative and thoughtful programming is possible.

The wide collection of case studies of practical experience from around the world is presented to help establish ways of working with communities experiencing great challenges. The book offers practical suggestions on how to give more substance to the rhetoric of community consultation and engagement in these areas of work. It suggests the need to work with a dynamic understanding of community formation that is particularly relevant when people experience unforeseen challenges and traumatic experiences. This title interrogates the concept of community through an extensive review of the literature and explores the ways of working with communities in transition and particularly in their recovery phases through an array of case studies in a range of socioeconomic and political contexts.

Focused on the concept of community in post-disaster recovery solutions—an aspect which has received little critical interrogation in the literature—this book will be a valuable resource to students and scholars in disaster management as well as humanitarian agencies.

Graham Marsh is a Visiting Research Fellow in Disaster Management at the School of Energy, Construction and Environment, Faculty of Engineering, Environment and Computing at Coventry University, UK.

Iftekhar Ahmed is a Senior Lecturer in the School of Architecture and Built Environment, University of Newcastle, Australia.

Martin Mulligan is an Associate Professor in the Sustainability and Urban Planning section of the School of Global, Urban and Social Studies at RMIT University and a former Director of RMIT's Globalism Research Centre, Australia.

Jenny Donovan is the Principal of Melbourne-based urban design practice Inclusive Design, Australia.

Steve Barton is a humanitarian training facilitator and freelance consultant working for agencies such as the Red Cross and the UN and is also founder and director of the Recovery Resource Centre.

Routledge Studies in Hazards, Disaster Risk and Climate Change

Series Editor: Ilan Kelman, Reader in Risk, Resilience and Global Health at the Institute for Risk and Disaster Reduction (IRDR) and the Institute for Global Health (IGH), University College London (UCL).

This series provides a forum for original and vibrant research. It offers contributions from each of these communities as well as innovative titles that examine the links between hazards, disasters and climate change, to bring these schools of thought closer together. This series promotes interdisciplinary scholarly work that is empirically and theoretically informed, with titles reflecting the wealth of research being undertaken in these diverse and exciting fields.

www.routledge.com/Routledge-Studies-in-Hazards-Disaster-Risk-and-Climate-Change/book-series/HDC

Published:

Climate Hazard Crises in Asian Societies and Environments
Edited by Troy Sternberg

The Institutionalisation of Disaster Risk Reduction
South Africa and Neoliberal Governmentality
Gideon van Riet

Understanding Climate Change through Gender Relations
Edited by Susan Buckingham and Virginie Le Masson

Climate Change and Urban Settlements
Mahendra Sethi

Community Engagement in Post-Disaster Recovery
Edited by Graham Marsh, Iftekhar Ahmed, Martin Mulligan,
Jenny Donovan and Steve Barton

COMMUNITY ENGAGEMENT IN POST-DISASTER RECOVERY

*Edited by Graham Marsh, Iftekhar Ahmed,
Martin Mulligan, Jenny Donovan
and Steve Barton*

Routledge
Taylor & Francis Group

LONDON AND NEW YORK

First published 2018
by Routledge
2 Park Square, Milton Park, Abingdon, Oxon OX14 4RN

and by Routledge
711 Third Avenue, New York, NY 10017

Routledge is an imprint of the Taylor & Francis Group, an informa business

British Library Cataloguing-in-Publication Data
A catalogue record for this book is available from the British Library

Library of Congress Cataloging-in-Publication Data
A catalog record for this book has been requested

ISBN: 978-1-138-69165-0 (hbk)
ISBN: 978-1-138-69167-4 (pbk)
ISBN: 978-1-315-53421-3 (ebk)

Typeset in Bembo
by Apex CoVantage, LLC

CONTENTS

FIGURES

TABLES

CONTRIBUTORS

Iftekhar Ahmed is a Senior Lecturer in the School of Architecture and Built Environment, University of Newcastle, Australia. Previously he was a Lecturer and Research Fellow in the School of Architecture & Design, RMIT University, Melbourne. His work focuses on resilient and sustainable post-disaster housing systems, disaster risk reduction and climate change adaptation. He previously also taught and conducted research at the University of Melbourne and Monash University, Australia. In the past, he has worked as a project manager at the Asian Disaster Preparedness Center (ADPC), Thailand, and as a shelter specialist at the United Nations Development Programme (UNDP), Bangladesh, and taught at the Bangladesh University of Engineering and Technology. Dr Ahmed has also served widely as a consultant for several international development and humanitarian agencies, particularly those working in the disaster risk reduction field. He has written a number of books and professional reports and has many peer-reviewed publications to his credit.

Steve Barton moved to the humanitarian and development sector after a long and varied career in design and construction. Two years as volunteer building schools on remote Pacific islands and a stint in Aboriginal community managing housing was his entry point to front-line disaster response and community development. He has been deployed in shelter, coordination refugee and programme management roles in Central Asia, Myanmar, Indonesia, Africa, and numerous locations in the Pacific, responding to earthquake, tsunami, cyclones, cold wave, malnutrition and conflict-based population movement. He has also designed and managed school and health centre building programmes. He is a humanitarian training facilitator and freelance consultant working for agencies such as the Red Cross and the UN and is also founder and director of the Recovery Resource Centre, an Australian NGO with the objective to develop bottom-up (beneficiary-driven) approaches in humanitarian and developmental response.

Amber Bennett is an award-winning communications and engagement strategist. She has honed her skills over 15 years, working with government, non-profits and business. She is the Principal of Upāya Consulting, an environmental communications consulting practice based in Calgary, Alberta. She holds a Master of Arts in Environmental Education and Communications from Royal Roads University and a Bachelor of Public Relations from Mount Saint Vincent University. She works primarily with local government, and her current projects include developing a climate change community mobilisation framework and an education and outreach framework to support riparian stewardship.

Beau B. Beza is Senior Lecturer in Deakin University's renowned Planning and Landscape Architecture Programme, School of Architecture and Built Environment. He co-edited and co-authored three chapters for *The Public City* (2014). He uses a place-making approach and an asset-based and co-production model to realise 'places' in urban and rural settings. Beau has worked on academic and consultancy projects in Australia, Bosnia-Herzegovina, Colombia, Nepal, Norway, Mexico, the Hashemite Kingdom of Jordan, and the US, where he completed his undergraduate university studies. Beau financially manages and is a team member in a $1.4m research project on improving the methods and impacts of agricultural extension in conflict areas of Mindanao in the Philippines.

Eva Bogdan has been studying and practising at the cross-section of society and environment—touching on topics such as floods, fuels, food and farming—around the world in Namibia, New Zealand, Europe and various parts of Canada. Broadly, she is interested in how diverse and potentially competing sets of values, viewpoints and interests are deliberated and decided upon in natural resource management and risk management. She is currently completing her PhD at the University of Alberta examining perceptions and practices of flood management in her province. In order to gain a more holistic understanding of the socio-political complexities in which environmental issues are embedded, she also obtained degrees in environmental science and sociology, as well as certificates in adult education, local government administration (NACLAA I), economic development, and community-based research and evaluation. Eva has worked on numerous research and community engagement projects, including co-designing and implementing a fun and interactive community disaster preparedness project in Alberta. For more information, visit www.evabogdan.com.

Sherri Brokopp Binder holds a PhD in Community and Cultural Psychology from the University of Hawai'i at Mānoa. Her research is focused on community resilience and disaster recovery, and includes qualitative and mixed-methods studies on the 2009 South Pacific tsunami in American Sāmoa and Hurricane Sandy. She holds an MA in Community and Cultural Psychology from the University of Hawai'i at Mānoa, an MA in Sustainable International Development from Brandeis University, and a BA in International Affairs from Kennesaw State University.

Sherri is currently a research consultant with Concept Systems, Inc., and a Commissioner on the Allentown Human Relations Commission in Allentown, PA. She previously served for seven years as the Sustainable Cities Division Director at the Urban Ecology Institute, an environmental policy institute at Boston College.

Yung-Fang Chen is a Senior Lecturer in Disaster Management and Emergency Planning and Director of the Centre for Disaster Management, Applied Research Group, Coventry University, UK. The focus of her research is on the methodology of planning and evaluating emergency response training and exercises. Her current projects include tasks for post-disaster assistance, in particular, shelter and housing, and community reconstruction. She is also involved in training university students to deliver disaster risk reduction training for primary school children. Other research interests include serious games for emergency response and e-learning pedagogy.

Jenny Donovan is the Principal of Melbourne-based urban design practice Inclusive Design. Her work spans the public, private and community sectors, and focuses on designing places that create the optimal chances for people to thrive and fulfil their potential. She is the author of *Designing to heal: Planning and urban design responses to disaster and conflict*, published by the CSIRO. The book is based on her experience of projects in Australia, the United Kingdom, Ireland, Ethiopia, Kosovo and Sri Lanka, and studies of projects in New York and Montserrat. She has been appointed an international expert in placemaking by UN Habitat.

Catherine Elliott has a PhD in the field of Geography and Environmental Studies from the University of Tasmania and works as a researcher at the University of Melbourne and as a Program Manager for the Komunitas untuk Hutan Sumatra (Community for Sumatra Nature Conservation). Her research includes residents' experiences of post-disaster housing in Aceh, Indonesia, following the 2004 Indian Ocean tsunami. She examines issues such as sense of place and dynamic processes of housing adaptation and advocates for an integrated approach to post-disaster housing that involves the intended residents in the planning process.

Anne Shangrila Y. Fuentes has worked on gender-related projects, such as integrating gender in the curriculum, and has taught courses on sex and culture, gender and sociology. Her research projects include sexuality education, gender, reproductive health, prostitution, hygiene behaviour of primary school children, indigenous concept of justice, extra-judicial killing and land conflict. She has provided trainings on gender-related topics among people in the academe, indigenous communities and villages within Mindanao. She also managed the gender and development activities of the University of the Philippines Mindanao when she was the Chair of the University Gender Committee. Her previous post as the Head of Office of the Tripartite Partnership among the Hiroshima University Partnership Project for Peacebuilding and Capacity Development (HiPeC), UP Mindanao and South–South Network provided her with the opportunity to deepen her knowledge on the various forms of conflict in Mindanao. She is currently involved in a four-year project, improving methods and impacts of agricultural extension in conflict areas of Mindanao, with the Australian Centre for International Agricultural Research and Royal Melbourne Institute of Technology.

Alex Greer is an Assistant Professor in the Political Science Department at Oklahoma State University. Alex conducts interdisciplinary, mixed-methods research on a number of elements of disaster science. Recent projects focused on: the relationship between issue framing and oil spill policy, household residential decision-making following disasters and the role that infrastructure plays in this process, archival disaster research on mental health response, and the development of a community resilience index. Alex has engaged in quick-response fieldwork after a number of events, including the Moore tornado of 2013, Hurricane Sandy in 2012, and the 2011 Tōhoku earthquake and tsunami.

Greg Ireton has worked in the post-disaster community and government nexus in Victoria for over 20 years. Working for both non-profit organisations and government, he has been involved in most of the significant disasters in Victoria over this period of time, as well as providing advice to other jurisdictions as they have undertaken significant recovery programmes. Following the 2009 bushfires, Greg was with the Victorian Bushfire Reconstruction and Recovery Authority, which provided further opportunities to work closely with the communities, local government and state government agencies. He has worked extensively within the state government and has had responsibility for a wide range of post-disaster recovery programmes. He is currently the Acting Manager of the Centre for Disaster Management and Public Safety as well as an Honorary Research Fellow within the Melbourne School of Population and Global Health, both within the University of Melbourne.

Mary Johnson is a research fellow with RMIT University, Australia, and chief investigator for an Australian government-funded project to improve methods and impacts of agricultural extension in conflict areas of Mindanao. Mary has worked extensively in agricultural, natural resource management, community development and education both within Australia and internationally. Community capacity building and developing strategic partnerships are critical for rural and remote communities to realise their own destinies, and Mary's experience in these areas has assisted in the support of vulnerable communities, including post-disaster and post-conflict. Recent consultancies have been in Uganda, Malawi and the Philippines and with Indigenous communities in Australia.

Lorenza Lazzati has completed a Master's in International Urban and Environmental Management at RMIT University, Australia, where she wrote a minor thesis on the contribution of participatory design to community resilience in post-earthquake situations in Italy. Since 2014 she has been collaborating with a Melbourne-based project management company to help develop their online community and post-disaster project management training. Their methodology bridges traditional project management with a community-based approach, putting stakeholders' values at the core of the project planning. The training has been delivered to several countries including Australia, Japan, Pakistan and the Philippines. Over the past two years, she has also been active in the non-profit sector and worked on the implementation of the Architects for Peace Strategic Plan and supported the organisation as a member of the Committee of Management. Lorenza has a background in architecture and she has worked in the construction sector for 15 years, developing broad experience on various projects, including hospitality, healthcare and commercial projects.

Marion E. MacLellan is a Senior Lecturer in Development in the Department of Geography, Environment and Disaster Management at Coventry University, UK. Her research interests lie predominantly in sub-Saharan Africa, and, in particular, in human development issues and disaster management. Her doctoral research focused on Rwanda and on the challenges faced by child-headed households in livelihood strategies and community integration. Her interest in conflicts and emergencies has led to her recently embarking on research into lost education in conflict and post-conflict societies, focusing on Sierra Leone and Rwanda. She has links with NGO projects in Mozambique, Rwanda and the Gambia, where she leads an undergraduate field trip each year. Her teaching includes responsibility for modules which focus on conflict, peacekeeping, food security and livelihoods, as well as more general development and disaster risk reduction.

Graham Marsh is a Visiting Research Fellow in Disaster Management at the School of Energy, Construction and Environment Faculty of Engineering, Environment and Computing. Previously he lectured and researched at RMIT and Coventry universities, specialising in recent decades in 'community' participation and particularly in relation to disaster management. He, along with colleagues, received a number of research grants from Emergency Management Australia to study community recovery and the reports were later published in the EMA's journal. Other consultancies included one covering 'Resilience for Scotland' for The Scottish Executive and others for the UNISDR. Graham was a member of the EU's COST Committee on Post Fire Forest Reconstruction across Southern Europe. He was also a consultant to the Project ForeStake research project of the University of Aveiro in Portugal, studying areas affected by forest fires in that country and the role of local stakeholders, and was a keynote speaker at their conference in 2013. Since 2007, he has worked with Coventry Master's graduates to establish Disaster Management Without Borders (DMSF) in the Democratic Republic of the Congo and a number of other countries across Africa.

Brendon McNiven is a principal at Global Engineering firm Arup. He is a structural engineer specialising in architectural buildings. He has worked around the world with Arup, and has been responsible for the delivery of many of the firm's significant projects, small and large. His specialist skills include lightweight structures, building stability systems, structural dynamics, façades, heritage and existing buildings, modular construction, emergency shelters, and BIM-assisted design delivery.

Martin Mulligan is an Associate Professor in the Sustainability and Urban Planning section of the School of Global, Urban and Social Studies at RMIT University and a former Director of RMIT's Globalism Research Centre. During his time in GRC, he was a lead investigator on a major research project looking at lessons to be learned from attempts to rebuild communities and household livelihoods in the wake of the 2004 tsunami disaster in Sri Lanka and southern India; a project which resulted in the publication of *Rebuilding Communities in the Wake of Disaster* (with Yaso Nadarajah; Routledge, 2012). In a wide range of research projects, he has looked at how local communities in Australia and Sri Lanka deal with 'external' challenges and the impact of globalisation and this has led him to undertake an extensive review of social, cultural and political theory on community formation in the contemporary world; resulting in an article entitled 'On Ambivalence and Hope in the Restless Search for Community' in the journal *Sociology* in 2014. His research is also reflected in *Introduction to Sustainability*, a textbook published by Earthscan/Routledge in 2015.

David O'Brien is a Senior Lecturer in the Faculty of Architecture, Building and Planning at the University of Melbourne. David coordinates the Bower Studio programme, which takes teams of Master's-level students to work alongside remote community groups to design and build infrastructure such as health clinics, education facilities, composting toilets and housing (see http://bowerstudio.com.au). David also conducts interdisciplinary research on post-disaster communities, with a special interest in settlement design and the ways residents modify their housing in the aftermath of the 2004 Indian Ocean earthquake and tsunami.

Mittul Vahanvati is the director of Melbourne-based architectural design and sustainability practice Giant Grass, and has both industry and academic experience. For more than seven years she has worked on low-tech, sustainable architecture, participatory approaches and hands-on learning. She is on design review panels at RMIT University and has co-authored the book *Co-building with Bamboo*. On the academic side, she has a Master's degree in sustainable development, and is currently pursuing a PhD in community resilience and post-disaster housing reconstruction approaches. Her recent paper on this subject was judged the best PhD research paper at the 5th International Conference on Building Resilience at the University of Newcastle, Australia. She has been a course coordinator, lecturer, tutor and researcher at RMIT, and other universities such as UNSW, UTS and Melbourne.

Her current work and research focus on rural/Indigenous cultures, ecologically sustainable shelter designs, disaster resilience and systems-based project life cycle management.

Lilia Yumagulova was born and raised in the Soviet Union, in a low-income area of a large urban centre prone to recurring floods. She holds degrees in engineering (Ufa State Aviation Technical University, Russia) and risk analysis (King's College London, UK). She is currently completing her PhD in Disaster Management Planning at UBC's School of Community and Regional Planning, focusing on regional disaster resilience planning. Her interdisciplinary academic path combines engineering, social science, public policy, international relations and planning. Throughout her academic career, Lilia has researched and worked in a variety of academic disciplines and institutions, including the University of Toronto, Canada; York University, Canada; University College London, UK; Lund University, Sweden; and UFZ, Germany, among other institutions. Lilia has worked in the media, government agencies and NGOs. In her academic work, she emphasises the value of bringing together academia, practitioners, policy makers, planners and media for reducing environmental risk in cities. Lilia is a Partnership Development and Research Co-Director for Crisis Resilience Alliance, Sauder School of Business, University of British Columbia. She is a Bullitt Foundation Fellow, Trudeau Foundation Scholar, Pacific Institute for Climate Solutions Fellow and Liu Institute for Global Issues Scholar. Her current work and research focus on urban and regional resilience planning, climate change adaptation, risk and crisis communication and management.

FOREWORD

John Twigg, Overseas Development Institute

This book makes an important contribution to our knowledge of two complex topics: disaster recovery and community engagement in disaster risk management. Both have been researched and written about before, but they are rarely examined together in this way.

Recovery is still under-researched compared to many other areas of disaster studies, and although new insights into recovery processes are emerging, the empirical evidence remains patchy. While there is plenty of writing on the impact of different approaches to reconstruction, for instance, there is relatively little on social dimensions of recovery. Community engagement in disaster-related actions has attracted a lot of research interest, but too much of that literature is uncritical and some is partisan.

We are still waiting for a more comprehensive, nuanced coverage of both of these topics, but reading this book helps to fill some of the gaps. It consists of empirical studies of disaster events and post-disaster actions: more empirical research is very welcome. The coverage is very broad, too, ranging across 13 countries with very different political, social and economic contexts; the book also covers different types and scales of disaster.

These studies reveal the remarkable diversity of 'recovery', which should make us more cautious about making general statements on the subject. They also tell us a great deal about recovery processes, actors and relationships, throwing light on the dilemmas, challenges and compromises that take place unavoidably in recovery initiatives.

The research also underlines the facts that communities are not homogeneous and that post-disaster recovery can be a challenge to community coherence. On the other hand, it can also lead to new or different forms of collaboration and partnership.

There are many lessons here for practitioners, policy makers and researchers. This book has certainly enriched my understanding!

PREFACE

Graham Marsh, Iftekhar Ahmed, Martin Mulligan,
Jenny Donovan and Steve Barton

The literature on disaster and/or risk management makes frequent references to community consultation, community engagement and community recovery, but the terms are used rather loosely, and there is little to suggest that the rhetoric is supported by analysis of practical experience in a diversity of contexts. To make matters worse, there tend to be high turnover rates for people engaged in disaster management and response and new people may have little opportunity to learn from past experiences. This is compounded by the diverse nature of disasters which are brought about by many different events that will have a unique impact and might occur anywhere on the surface of the earth. The specialised nature of those engaged in disaster management often means moving to address the latest disaster and bringing values and experiences gained from different disasters to their most recent projects. These past experiences are not always relevant and the application of the lessons learnt can indeed be counter-productive.

The origins of this book lie in the editors grappling with what is seen to be a major issue for those involved in the rebuilding processes following the results of a disaster or conflict: the 'complexities' associated with 'community', particularly in relation to disaster recovery, and the various ways in which those involved in emergency response and rebuilding engaged, or failed to engage, with the community in the processes.

Each of these editors perceived that there are gaps in the literature with regard to 'community'. Gaps which stem from there being such a variety of definitions of 'community' and what can be portrayed as a general lack of understanding of what a 'community' is. In fact, the concept of 'community' means different things for different people in different circumstances.

Questions arose through previous research and work of the editors. How does one define 'community'? How, if at all, is it defined/perceived by practitioners in the various fields? Does 'community' even exist? Can we say there is no such thing as '*the* community'? How is it (assuming 'it' or some such entity exists) affected by a disaster? In the field, the concept certainly is that it is present, but what exactly is it in reality? Is it generally the case that in any geographic area there may be many communities of interest, which may be in conflict with each other? Too often, and particularly in war-torn countries, conflict or distrust may be the norm due to past history. Can it be the situation, perhaps due to internal immigration, due to previous disasters, conflict or economic circumstances, that there are almost no relationships between residents in the locality and those that do exist may be patchy at best?

All these were questions raised by the editors and by many researchers and practitioners in the field of disaster management and they led to a number of conclusions. Diversity and 'community' do go together. There is no one way to deal with 'community'. What works in one situation may not be as effective in another.

What should be aimed for in the recovery processes is how to achieve the best outcome for the maximum number of people.

The first chapter begins with an outline of the meaning of community in the contemporary world, which is based on an extensive review of the sociological literature on the topic. This suggests a need to work with a dynamic understanding of community formation that is particularly relevant when people experience unforeseen challenges and traumatic experiences.

In the subsequent chapters, authors from across the globe use case studies of both practical experience and research to examine ways of working with communities in transition or trauma, and particularly in their recovery phases. In this way, the book as a whole can offer practical suggestions on how to give more substance to the rhetoric of community consultation and engagement in these areas of work. The chapters are written by both academics and practitioners, in the hope that the findings in this book will be of use in research and programme implementation.

1

RETHINKING THE MEANING OF 'COMMUNITY' IN COMMUNITY-BASED DISASTER MANAGEMENT

Martin Mulligan

Introduction

The need to consult or engage vulnerable or disaster-affected communities in disaster mitigation or recovery work is a core assumption for those who work in this field of practice. It is included in most national disaster management strategies around the world and features in the policy commitments of government and non-government international relief and aid agencies. The lead author for the United Nations Office for Disaster Risk Reduction's 2013 Global Assessment Report, Andrew Maskrey, has argued that the need to engage local communities in disaster management has been repeatedly confirmed since it was first mooted at an international conference held in 1984.[1] The key advocates of community-based disaster management (CBDM) focus primarily on risk reduction and disaster avoidance, but they note that the importance of community engagement is highlighted by experiences of disasters and community participation policies are put to the test in disaster recovery operations. Numerous studies have urged the need for community participation in disaster recovery work.[2]

However, many researchers have also noted that post-disaster community participation practices frequently fall short of the policy rhetoric,[3] because it is much more difficult to work with complex, traumatised local communities than most practitioners imagine and very few of them have ever had any training in this work. Apart from the fact that the difficulties involved in engaging disaster-affected communities in recovery work is frequently underestimated, the problems stem from the fact that the word 'community' is used very loosely and uncritically in policy and practice literature. The single word is commonly used to refer to local or place-based communities, rather than multiple and overlapping forms of community, and it is often assumed that the pre-disaster local communities were relatively stable and homogeneous rather than dynamic and multifaceted, to the point of being internally conflicted at times. Natural disasters can open up 'fault-lines' within local communities related to class, religion, ethnicity and gender,[4] and relief and aid agencies can unwittingly exacerbate divisions and tensions by acting without adequate knowledge of the local complexities.[5] Given the frequency with which the word 'community' is deployed in disaster management literature, there has been surprisingly little commentary on how the word can be unintentionally misused or even intentionally abused by actors with partisan or self-serving agendas.

As Raymond Williams noted in 1983, the word 'community' is a 'keyword' in the English language, because it touches on an eternal desire to find a sense of belonging in an ever-changing world, and similar words crop up in other human languages. However, as Williams also noted, it can mean very different things to different people, and those who want to mobilise community in order to increase social cohesion and adaptive capacity need to use the word thoughtfully and carefully. In his seminal work on the sociology of community,[6] Gerard Delanty noted that new communication technologies mean that individuals can belong to many more forms of 'real' and 'virtual' communities than ever before. However, the stretching of what the word refers to does not mean that it has become any less relevant; Delanty (2003) argued that 'the persistence of community consists in its ability to communicate ways of belonging, especially in the context of an increasingly insecure world.'[7] Further, he argued:

> Community is relevant today because, on the one side, the fragmentation of society has provoked a worldwide search for community, and on the other . . . cultural developments and global forms of communication have facilitated the construction of community.[8]

Sociologists have agonised over the meaning and relevance of the idea of 'community' since a celebrated debate on the topic between founding fathers Ferdinand Tönnies and Émile Durkheim over a century ago.[9] Delanty and others[10] have countered those who have suggested that 'urbanisation' and 'globalisation' have rendered the idea of community obsolete. However, community formations have become more multilayered and impermanent than ever before—in all human societies—and, consequently, the unexpected consequences emanating from simplistic or uncritical use of the word have intensified. Other chapters in this book will demonstrate that many disaster management practitioners find ways to work effectively within complex and traumatised local communities. However, sophisticated understandings of community are yet to emerge in disaster management literature, perhaps because practices have not been adequately informed by insights drawn from sociology and anthropology.

The symbolic importance of community

In his debate with Tönnies, Durkheim suggested that steady and accelerating migration of people from rural villages into burgeoning cities—a process which started in Europe but which has spread globally in more recent times—may have made the concept of community obsolete. The US-based 'Chicago School' of sociologists subsequently argued that apparent differences between Durkheim and Tönnies on the lingering importance of old forms of community could be resolved because distinct local communities emerge with the 'mosaic of little worlds'[11] taking shape within cities. This helped to emphasise the ways in which communities can emerge within different geographical and social contexts. However, the anthropologist Anthony Cohen pointed out[12] that studies inspired by the Chicago School presented a rather mechanical rendition of community formation, which did not take into account Durkheim's interesting distinction between 'mechanical' and 'organic' forms of solidarity. According to Cohen, an emphasis on the mechanics of community formation fails to take into account the deep and eternal human need to find a sense of belonging to community, and Chicago School leaders ended up painting a rather bleak picture of the ability of local urban communities to withstand 'external' economic and social change.

In shifting the emphasis to the enduring symbolic importance of community, Cohen argued that the desire to experience community is rarely extinguished, even in the most challenging circumstances. As an anthropologist, he drew on the work of his mentor Victor Turner on the topic of 'liminality',[13] which can be understood as moments evoked by ritualistic practices when normality is suspended.[14] Turner had argued that community exists 'in resistance to structure, at the edges of structure, and from beneath structure'[15] and Cohen cited this in order to argue that the time had come for sociologists to shift from debates about the 'lexical meaning' of community to emergent understandings of community formation and constant reformation. According to Cohen, it is also important to maintain a dynamic understanding of how communities construct boundaries to distinguish themselves from other communities. In Cohen's approach, a community is not formed through the depiction of a territory but by clustering around symbols of a shared identity, even if geography sometimes suggests the ways in which one community might distinguish itself from another. However, he makes the point that any boundary will enclose difference or diversity as much as sameness and the whole process of boundary-setting can open up discussions and debates about identity and belonging.

Impermanence and incompletion of community

If we shift from thinking of community as a social structure to see it as a largely irrepressible human desire, we need to acknowledge that the desire can never actually be fulfilled. This idea about the essential 'incompletion' of community has probably been best expressed by the French philosopher and art critic Jean-Luc Nancy in his famous essay titled *The Inoperative Community*, which was first published as an article in 1983 before being re-published in book form in 1991.[16] According to Nancy, people are most likely to think about the importance of community when they feel it has somehow been lost or is missing, and this certainly applies to people who have experienced a natural disaster. This emphasis on the fragility of community makes it clear that it can never be taken for granted and Nancy went on to argue that 'incompletion' is the necessary 'principle' of community. Delanty has warned that the principle of incompletion can represent community as little more than a cruel illusion, and use of the word 'inoperative' may have overstated the argument, for Nancy himself stressed the need to think of 'incompletion in an active sense . . . as designating not an insufficiency or lack, but the activity of sharing'.[17]

Delanty extended the argument about the incompletion of community by noting that it always involves an interplay between individualism and solidarity. Invariably, he argued, community 'ends up [being] destroyed by the individualism that created the desire for it'.[18] Nevertheless, the search for community continues because it 'offers people what neither society nor the state can offer, namely a sense of belonging in an insecure world'.[19] In making these points, Delanty echoes earlier work by Nikolas Rose who argued for the need to think of community as 'localized, fragmented, hybrid, multiple, overlapping and activated differently in different arenas and practices'.[20] In the contemporary world, Rose stressed, 'individuals no longer inhabit a single "public sphere"', and this means that 'communities can be imagined and enacted as mobile collectivities, as spaces of indeterminacy, of becoming'.[21] We need to think of community as 'not fixed and given but locally and situationally constructed'.[22] 'To community as essence, origin, fixity, one can thus counterpose community as a constructed form for the collective unworking of identities and moralities.'[23]

The 'principle' of incompletion invokes the 'principle of impermanence', which is shared by ontologies as different as Buddhism and complexity science. While the work cited above on the sociology of community in the contemporary world comes from western scholarship, it suggests that the constant search for community is a fundamental human need or aspiration. Economic, social, technical and cultural dimensions of 'globalisation' have simultaneously removed barriers of distinction between the ways in which societies operate in different parts of the world and introduced new opportunities for participation in layers of social formation. The increased mobility of people, goods, information and ideas means that most local communities are more fluid—or impermanent—than ever before and individuals living in settings as diverse as rural Nepal and urban England can belong to a host of real and virtual (communicatively constructed) communities. Local communities are more likely than ever to include a host of 'sub-communities' that can reach beyond their boundaries, including diasporic communities, religious or ethnic communities, and a wide variety of interest groups. Allegiances to non-local forms of community can make local coexistence more problematic, as seen in conflict-ridden nations ranging from Syria to Sri Lanka. Natural disasters can hit communities that are far from being stable or homogeneous and, like it or not, relief and aid workers can either exacerbate or reduce local tensions and divisions.

The twin principles of incompletion and impermanence make it clear that the existence of community can never be taken as a given and Delanty hammered this point by saying that community only exists in the contemporary world to the extent to which it is 'wilfully constructed'.[24] This suggests that relief and aid workers do not encounter pre-existing communities but rather communities which may be coming in and out of existence and they will play a role in the construction or deconstruction of community, even if this can only be by aiding and abetting the activities of community members.

The dark side of community

While a more nuanced understanding of community formation is emerging within varying fields of practice and scholarship, it is important to take heed of persistent warnings that community is not always the wholesome experience that it promises to be. A projection of community identity will commonly create a sense of there being 'insiders' and 'outsiders', and those who feel excluded will feel that exclusion all the more for the fact that others feel included. 'Security for some', wrote Mae Shaw,[25] 'may be achieved only by the exclusion of others', as seen in the exclusion of asylum seekers from citizenship in most western societies. Even worse, Iris Marion Young noted[26] that projections of community identities can often be bound up with ethnic, racial and class divisions and related feelings of either superiority or resentment. Indeed, communalism born of resentment can lead to sometimes sudden and unexpected outbreaks of communal violence.

Eruptions of communal violence can be made worse when tensions and resentments simmer from one generation to the next, and the Indian sociologist Manoj Jha[27] has provided an insightful account of how this played out in the alarmingly brutal attacks on Muslims living in the Indian state of Gujarat over a period of four months in 2002. Those responsible for the violence tried to characterise it as justifiable revenge for past injustices, and Jha suggested that communities sometimes draw on stories of past humiliation as the 'chosen traumas' to justify revenge. Things that may have happened in the distant past can be 'psychologised' and 'mythologised' to become 'markers of their identity', Jha wrote.[28] 'Once a trauma becomes a "chosen trauma" the historical truth about it does not really matter'[29] and this makes it difficult to subject the conflict to 'rational' analysis. In such circumstances, Jha suggested, local authorities and community workers need to seek ways for the 'strengthening of shared spaces, shared interests and shared destiny', allowing for 'points of encounter' between 'open expression of a painful past' and the 'articulation of a long-term independent future'.[30] Ultimately, Jha concluded, the aim is to build trust among communities 'who have either forgotten the beauty of plural living or have fallen prey to . . . manufactured amnesia'.[31]

Inter-communal conflict has a long history in many societies across the world, but the communal riots that broke out across the UK in August 2011 suggest that there is potential for sudden upsurges of communal violence in any society. This reminds us that the very idea of community can be emotionally fraught and hotly contested, and here it is useful to turn to Roberto Esposito's work on the 'origins and destiny of community'.[32] Noting that many European languages have a version of the word 'community' which emanates from the Latin word *communitas*, Esposito suggests that the Latin word involves a combination of the word *commun* to refer to that which 'begins where what is proper ends'[33] and the word *munus* which is, in turn, linked to notions of gift and obligation.[34] It is interesting to note that the word *commun* appears in Greek as *koinos* and in German as *gemein*,[35] and that the opposite of *munus* is *immunus*, which translates loosely as immunity to exchange relationships that trigger obligations.[36] According to Esposito's analysis, the *munus* that comes to be shared publicly is not 'a property' or 'possession' but rather 'a debt, a pledge, a gift that is to be given'.[37] This raises the idea that the sense of obligation that is embedded within the word *communitas* refers to a debt or a 'lack', rather than something that already exists. This innovative interpretation of the sense of obligation embedded within the word 'community' leads Esposito to make the following incisive observation:

> Seen from this point of view, therefore, community isn't only to be identified with the *rea publica*, with the common 'thing', but rather it is the hole into which the common thing continually risks falling, a sort of landslide produced laterally and within. This fault-line that surrounds and penetrates the 'social' is always perceived as the constitute danger of our co-living . . . We need to watch out for this without forgetting that it is *communitas* itself that causes the landslide; the threshold that we can't leave behind because it always outruns us . . . as the unreachable Object into which our subjectivity risks falling and being lost. Here then is the blinding truth that is kept within the etymological folds of *communitas*; the public thing [*rea publica*] is inseparable from the no-thing [*niente*].[38]

Here we hear echoes of Nancy's work on the incompletion of community and his emphasis on experiences of community through loss or lack. However, it presents an even more frightening prospect of falling into a void instead of finding the safety of community. It suggests that we are driven by a combination of hope and fear in reaching for something that we can never fully grasp. This helps to explain why the idea of community has persistent and powerful emotive appeal for people everywhere, but it also serves to remind us that there is a fine line between hope and despair.

Learning from practice

While the core principles of community-based disaster management have been embraced by international relief and aid agencies,[39] there is little evidence to suggest that the lessons of good practice are being documented and disseminated. There is also little evidence to suggest that the theories and practices of community development that have emerged in countries such as the UK, USA and Australia over more than 40 years are used to prepare people working in community-based disaster management internationally. Specialist journals such as *Disasters* and the international *Community Development Journal* appear to operate in different and disconnected domains. While papers in *Disasters* rarely discuss the complexities of community participation or engagement, papers in *Community Development Journal* commonly explore the dilemmas faced by practitioners in trying to advocate for the needs and aspirations of vulnerable communities.

In a book marking the 40th anniversary of the launching of *CDJ*, Gary Craig notes that 1968 was a turning point for the emergence of a practice driven by the rhetoric of 'empowerment', which was, in turn, inspired by the 'community education' pedagogies of Paolo Freire and Ivan Illich.[40] While the rhetoric promised 'liberation' from dependence on state welfare, the state detected an opportunity to redirect its welfare spending and, as Craig notes, 1968 also saw the birth of Britain's Community Development Project (CDP) which offered funding for community development work. Hence, the field of practice that spread across a wide range of western nations in 1970s had a rather paradoxical relationship with the state. Nikolas Rose detects a shift from seeing community as 'critique and opposition directed against remote bureaucracy'[41] to an 'expert discourse and a professional vocation', which brought with it the 'suspicion' that 'the space of community was being colonized by agents, institutions and practices of control'.[42] Perhaps the suggestion that community development became, at least potentially, a new form of government control helps to explain why it has been neglected by sociologists; although a more benign explanation may be that it was seen as falling within the category of 'social work' rather than 'sociology'. However, according to Mae Shaw,[43] community development can also be seen as a new way of engaging with the state, in which those who advocate for 'problem communities' can force the state to become more responsive. According to community development scholar Marjorie Mayo,[44] efforts to empower local communities have not only been confronted by the ambiguities embedded in the word 'community', but also by the fact that 'it has been contested, fought over and appropriated for different uses and interests to justify different politics, policies and practices'.

While 'community development' practice may have emerged first in countries such as the UK, USA and Australia, the papers published in *CDJ* for more than four decades make it clear that it has much in common with sometimes older practices in countries such as India, Brazil and the Philippines. It has clearly emerged as a distinct international field of practice and yet it does not attract the attention it deserves, even from those who embrace the rhetoric of community participation in disaster management work. The practice of community development has not informed scholarly debates on the changing meaning of 'community' to the extent it could, and more needs to be done to bridge the gap between theory and practice.

One person who made a valiant effort to bridge this gap was a very experienced urban community development worker in England, Jeremy Brent, who brought 28 years of practical experience into his late career university studies. Although he died in 2006, a group of friends and admirers decided to turn his PhD thesis into a book titled *Searching for Community: Representation, Power and Action on an Urban*

Estate, published in 2009. The thesis/book ends with a series of reflections drawn from Brent's study and experience and it concludes with the observation: 'Community is not a simple concept and is dangerous if it is simplified.' 'Engaging with community', he went on to say, 'is a practice full of ambivalence, but always one full of hope.'[45]

Notes

1 See Maskrey 2011, pp. 42–52.
2 See, for example, Barenstein and Leemann 2012; Kenny 2010; Lyons *et al.* 2010; Lizzeralde *et al.* 2009.
3 See, for example, Cosgrave 2007; Mulligan and Nadarajah 2012; Barenstein and Leemann 2012.
4 See Mulligan 2013.
5 Ibid.
6 The first edition of Delanty's book on community was published in 2003 and a second edition appeared in 2010.
7 Delanty 2003, p. 187.
8 Delanty 2003, p. 193.
9 See Aldous *et al.* 1972.
10 Including, for example, Nikolas Rose (1996), Zygmunt Bauman (2001) and Richard Sennett.
11 This term comes from Chicago School leader Robert Park in a paper in an American sociology journal in 1915.
12 Cohen 1985.
13 See Turner 1969.
14 Delanty 2003, p. 44.
15 Turner 1969, p. 128.
16 Others to express similar ideas about community were French sociologists Maurice Blanchot and Georges Bataille, cited by Delanty 2003, p. 135.
17 Nancy 1991, p. 35.
18 Delanty 2003, p. 192.
19 Ibid., p. 195.
20 Rose 1999, p. 178.
21 Ibid., p. 195.
22 Ibid.
23 Ibid.
24 Delanty 2003, p. 130.
25 Shaw 2007, p. 28.
26 Young 1990.
27 Jha 2010.
28 Ibid., p. 318.
29 Ibid.
30 Ibid.
31 Ibid.
32 Esposito 2010.
33 Ibid., p. 3.
34 Ibid., p. 4.
35 Ibid.
36 Ibid., pp 4–6.
37 Ibid., p. 6.
38 Ibid., p. 8.

39 Maskrey 2011, p. 46.
40 Craig 2008, p. 182.
41 Rose 1999, p. 175.
42 Ibid., p. 176.
43 Shaw 2007.
44 Mayo 2008, p. 48.
45 Brent 2009, p. 261.

References

Aldous, Joan, Émile Durkheim and Ferdinand Tönnies, 1972, 'An Exchange between Durkheim and Tönnies on the Nature of Social Relations with an Introduction by Joan Aldous', *American Journal of Sociology* 77:6, pp. 1191–1200.

Barenstein, Jennifer and Esther Leemann (eds), 2012, *Post Disaster Reconstruction and Change: Communities' Perspectives*, Boca Raton, FL: CRC Press.

Bauman, Zygmunt, 2001, *Community: Seeking Safety in an Insecure World*, Cambridge: Polity Press.

Brent, Jeremy, 2009, *Searching for Community: Representation, Power and Action on an Urban Housing Estate*, Bristol: Polity Press.

Cohen, Anthony, 1985, *The Symbolic Construction of Community*, London: Tavistock.

Cosgrave, John, 2007, *Synthesis Report: Expanded Summary, Joint Evaluation of the International Response to the Indian Ocean Tsunami*, London: Tsunami Evaluation Centre.

Craig, Gary, 2008, 'Community Work and the State', in Gary Craig, Keith Popple and Mae Shaw (eds) *Community Development in Theory and Practice: An International Reader*, Nottingham: Spokesman Books.

Delanty, Gerard, 2003, *Community*, London: Routledge.

Esposito, Roberto, 2010, *Communitas: The Origin and Destiny of Community*, Stanford, CA: Stanford University Press.

Jha, Manoj, 2010, 'Community Organization in Split Societies', *Community Development Journal* 44:3, pp. 305–319.

Kenny, Sue, 2010, 'Reconstruction through Participatory Practice?' in M. Clark, I. Fanany and S. Kenny (eds) *Post-Disaster Reconstruction: Lessons from Aceh*, pp. 79–106, London: Earthscan.

Lizzeralde, G., C. Johnson and C. Davidson, 2009, *Rebuilding after Disasters*, London: Routledge.

Lyons, M., 2010, 'Can Large-Scale Participation Be People-Centred? Evaluating Reconstruction as Development', in M. Lyons, T. Schilderman and C. Boano (eds) *Building Back Better: Delivering People-Centred Housing Reconstruction at Scale*, Schumacher Centre for Technology and Development, Warwickshire: Practical Action Publishing.

Maskrey, A., 2011, 'Revisiting Community-Based Disaster Risk Management', *Environmental Hazards* 10, pp. 42–52. Available at: www.tandfonline.com/loi/tenh20.

Mayo, Marjorie, 2008, 'Community Development: Contestations, Continuities and Change', in Gary Craig, Keith Popple and Mae Shaw (eds) *Community Development in Theory and Practice: An International Reader*, Nottingham: Spokesman Books.

Mulligan, Martin, 2013, 'Rebuilding Communities after Disasters: Lessons from the Tsunami Disaster in Sri Lanka', *Global Policy* 4:3, pp. 278–287.

Mulligan, Martin and Yaso Nadarajah, 2012, *Rebuilding Communities in the Wake of Disaster: Social Recovery in Sri Lanka and India*, New Delhi: Routledge.

Nancy, Jean-Luc, 1991, *The Inoperative Community*, Minneapolis: University of Minnesota Press.

Park, Robert, 1915, 'The City: Suggestions for the Investigation of Human Behaviour in the City', *American Journal of Sociology* 20, pp. 577–612.

Rose, Nikolas, 1996, 'The Death of the Social', *Economy and Society* 25, pp. 327–356.

Rose, Nikolas, 1999, *Powers of Freedom: Reframing Political Thought*, Cambridge: Cambridge University Press.

Shaw, Mae, 2007, 'Community Development and the Politics of Community', *Community Development Journal* 43:1, pp. 24–36.

Turner, Victor, 1969, *The Ritual Process: Structure and Anti-Structure*, London: Routledge.

Williams, Raymond, 1983, *Keywords: A Vocabulary of Culture and Society*, London: Flamingo.

Young, Iris Marion, 1990, 'The Ideal of Community and the Politics of Difference', in L.J. Nicholson (ed.) *Feminism/Postmodernism*, New York: Routledge.

2

REBUILDING LESSONS FROM BUSHFIRE-AFFECTED COMMUNITIES IN VICTORIA, AUSTRALIA

Greg Ireton and Iftekhar Ahmed

Introduction

This paper explores the complexity and challenges of post disaster rebuilding through people's lived experience of the 'Black Saturday' bushfires of 2009 in the state of Victoria, Australia. It outlines the tension between deliberative and speedy rebuilding, the contextual issues that impact on decision-making and discusses some of the process found to be successful in assisting people in their decision to rebuild. The paper then outlines a number of gaps and recommendations for policy-makers and practitioners.

The 2009 bushfires in Victoria saw the greatest loss of life in any bushfire in Australian history. There were 173 fatalities, more than 2,300 homes were destroyed, 43,000 hectares of land were burnt, more than 70 national parks and reserves were damaged, and 11,000 farm animals were killed or injured. More than 10,000 insurance claims were made, totalling AUD$1.09 billion (VBRRA, 2011).

Post-disaster housing reconstruction in Australia, as in many developed countries, is owner-driven. Private resources such as insurance, loans and savings usually provide the primary means for funding repair and rebuilding (Zhang and Peacock, 2010). The government's role is primarily to provide information, advice and short-term and temporary accommodation, as well as the provision of limited financial assistance based on eligibility criteria. The range of activities undertaken by the state government was expanded due to the scale of the 2009 bushfires, and additional assistance that became available through the Victorian Bushfire Appeal Fund.

Catastrophic disasters, such as the 2009 bushfires, have a wide range of impacts that persist over a long time and these events often illustrate the challenges of post-disaster housing policy and programmes. Sapat *et al.* (2011, p. 26) found that, "As a policy issue, post-disaster housing continues to be a 'wicked' and 'messy' policy problem, exacerbated by unrealistic expectations of governmental agencies." For catastrophic events in particular, it is clear that housing recovery is not a short-term activity. The issue cannot be left solely to the market and must be incorporated as a core part of the long-term recovery planning (Zhang and Peacock, 2010).

Well-considered and deliberative decision-making is often challenging in the post-disaster environment when there is pressure to rebuild quickly (Evans-Cowley and Kitchen, 2011; Olshansky, 2006; Olshansky *et al.*, 2008; Paul and Che, 2011; Zhang and Peacock, 2010). The sudden change in the context is often not fully understood by institutions and communities until investigations are completed and responses decided upon. This can be exacerbated by further uncertainty arising through planned or potential changes to land planning, building regulations and inquiry processes. These are often flagged in the weeks and months following many large-scale disaster events. Many people are keen to rebuild quickly, based on the pre-existing housing and infrastructure to restore a familiar community, while others advocate reconstruction that incorporates new planning and disaster-risk reduction. A tension often develops between quickly rebuilding and re-establishing known social and economic networks versus a slower and more deliberative process that focuses on long-term community development outcomes. The impact of this tension becomes apparent when people reflect on the importance of what appeared to be minor issues at the time of the rebuilding process. One example was provided following the 2009 bushfires, when the Victorian Government provided a free clean-up programme to expedite the often lengthy process. Although the people-centred approach used by the lead contractor was typically well regarded by participants, the speed of the clean-up did occasionally lead to challenges. At a community leadership forum in 2010, an account from a member of the bushfire-affected community in Christmas Hill highlights the importance of seemingly minor matters:

> Our insurance company was efficient. Too efficient. The demolition crew arrived unannounced. Before the government had organised [the developer] clean-up and before we had a chance to search properly for surviving 'treasures', my husband discovered the demolition activity by accident, having driven past our property for a quick look. He threw himself in front of the bulldozer to stop them and called me to bring the kids so we could have a quick sift through before they continued.

Stories such as this clearly indicate that a process focusing not only on a speedy rebuild but also on individuals' emotional needs can play a very important part in contributing to longer-term recovery and well-being, beyond just the physical outcome of a house to inhabit.

Following a rapid rebuild, too often people lament that they are now living in a "house" and not the "home" which they lost. This differentiation between house and home is very personal and individual, but time to consider what might make the difference between a house and a home during the recovery process might assist people managing their expectations about what a rebuild can achieve.

It often takes time to make a house a home through the processes of habitation, personalisation and familiarity.

Lack of evidence-based information

These challenges in the post-disaster context are exacerbated by the lack of detailed studies and information on appropriate processes for rebuilding housing and recovering from the devastating social and physical losses that arise (Zhang and Peacock, 2010). In the aftermath of the 2009 bushfires, there was very limited information on how long the rebuilding process following a catastrophic bushfire might actually take, or on how many people would be likely to rebuild. Additionally, there was very little information on whether outcomes for individuals and families are improved if they remain within the community or move to another less affected community.

The dilemmas of rebuilding

Considering the level of devastation, loss of infrastructure and services, and lack of immediate access to services, relocation to less affected areas can seem like a beneficial option. It is equally easy to appreciate the benefits of remaining within a tight-knit community and maintaining existing social networks, with easy access and proximity to rebuilding sites (Bonnanno et al., 2010). In reality, the benefits and constraints in either of these choices depend on a range of factors within the affected community and surrounding areas. Such factors include the services available and timelines for personal financial resources, pre-existing issues or resources within the community prior to the disaster, disruption and likely restoration of lifelines and services, disruption to community networks, availability and access to employment, and the availability of housing.

Disaster-affected people often experience pressure to rebuild, be it real or perceived, from friends, donors, the government and the media. Many conversations with people considering their options after losing their home revealed that they felt they would be letting down the community and the broader society that had donated money if they did not rebuild within the community; their own or families' interest seemed of less importance. They experienced pressure to quickly make decisions that would impact their lives for a very long time.

A pre-conceived idea that often prevails within government and the media is that anyone who has lost a house to a bushfire (or other disaster) will naturally want to rebuild. There seems to be little thought about whether these people have prior experience in building a house, or have ever had the need or desire to build a house. This is not limited to bushfires in Australia. A resident in Kansas, USA, whose house was lost in a tornado in 2007, stated:

> With stunned minds, we began trying to decide what to do next. Ray did not have the mental and emotional energy to rebuild. I clung to the thought

of rebuilding for a while, but to be honest, the last thing that I'd ever wanted to do was build a house.

(Paul and Che, 2011, p. 101)

A common story in many of the bushfire-affected communities was of elderly couples who had painstakingly established extensive gardens over decades and had no inclination, and perhaps no longer the physical strength, to regrow their gardens. There are similar reports, some from bereaved people who did not want to consider rebuilding on the site where loved ones had perished; others who thought that they would never regain the lost sense of safety and security within their community. As well as this, a number of other people who had tentative plans to move out of the community prior to the bushfires saw this as an opportunity to act upon those plans.

Strategies for "holding the space"

The Victorian Bushfire Reconstruction and Recovery Authority (VBRRA) was formed after the 2009 bushfires by gathering together professionals with relevant skills from different government departments and other organisations. VBRRA took on a significant role in mediating between the affected communities and other government agencies, often advocating the "holding the space" approach, as aptly described by a community worker (River, 2013). This approach was applied first to ensure that immediate needs were addressed, such as short-term or temporary accommodation, access to support and advice, and immediate financial assistance where required. Second, the approach aimed to provide time and information to allow an individual or household to make informed decisions about what might be most beneficial for them.

Making informed decisions requires having access to a wide range of information and advice. Legal, insurance, and financial advice was made available directly through specialist advisers or pro-bono services facilitated through VBRRA. However, it was often found difficult to convince people of the importance of this advice prior to making rebuilding decisions. VBRRA's Rebuilding Advisory Services (RAS) officials, deployed after the bushfires, remarked at times despairingly that they were often asked for advice only after people fell into trouble resulting from a previous decision.

VBRRA supported and facilitated three main "holding the space" strategies: the development of an Urban Design Framework for Marysville; the building of temporary villages in a number of communities; and the provision of wider support during the transition to recovery, as discussed below.

Urban Design Framework

While considering whether to rebuild individually, people were also concerned about the nature of their community into the future and the opportunities presented in rebuilding that may incorporate improved safety and environmental, community, aesthetic or economic outcomes. VBRRA engaged urban planners to work with the local council and the community to establish a shared vision for Marysville, one of the most severely affected towns. A workshop was conducted to develop an Urban Design Framework for the community and surrounding townships. This enabled residents to be involved actively in shaping the future nature of their town and, through discussion, to gain a better understanding of the opportunities and alternatives. The involvement of urban designers in the workshop allowed the ideas of the participants to be sketched as streetscapes, better enabling the visualisation of the ideas as they became incorporated into the vision for the town's future. Drawings such as in Figure 2.1 were used for sparking conversations on what people liked or did not like in order to guide the planning process.

Temporary accommodation

The Victorian Government through VBRRA decided to build several temporary villages to provide housing options within some of the most severely affected communities. This was not part of any pre-existing planning and was developed initially in response to the desire of the bushfire-affected community of Flowerdale to have such a village (VBRRA, 2011).

Through further community consultations, two more temporary villages were built in Marysville and Kinglake (see Figure 2.2), and units were also built in Whittlesea. At their full capacity in April 2010, 314 people lived in the temporary villages, and the final residents moved out of the villages gradually during 2011–12. The pressure to build the villages quickly allowed little time to consider

FIGURE 2.1 Illustration showing a future vision for Marysville after rebuilding.
Source: Robert Days, 2011.

the most suitable or cost-effective solutions or what would be in the best interest of the community in the long term. This was particularly challenging because, as stated above, the villages were quickly established without any prior planning. It was also difficult to know how many people would eventually use the villages and how long they would stay there, requiring planning for expansion where limited suitable land was available for the purpose. Despite these challenges, a key success factor of the villages was that they prevented the community from disintegrating and dispersing. For many of the people who lived there, it was a positive experience, as captured by a resident's comment: "It [the temporary village] was the best ever thing that happened after the bushfire. It really kept our community together" (Charlesworth and Ahmed, 2015, p. 23).

Temporary accommodation cannot be successful in isolation and, to be effective, requires other forms of support and the active engagement of residents and the community. Thus, a range of communal facilities and activities was included in the temporary villages. RAS staff worked with the temporary village residents to develop housing plans for their permanent accommodation. Regular meetings with residents ensured that critical issues could be addressed by including ideas from them, such as to install backyards and pet enclosures, television antennas, wireless internet and a games room in the Marysville and Kinglake villages.

FIGURE 2.2 Aerial view of the temporary village in Kinglake.
Source: Max Ginn, 2010.

Wider support

For people remaining on their property, a range of government-facilitated initiatives sought to provide assistance. These included:

- A clean-up programme that cleared a total of 3,053 properties, with 98 per cent of these properties cleared within 18 weeks of the fire. The programme was designed to meet the community's health and safety requirements, as well as support those who required demolition and clean-up services. The coordinated service reduced the need for professional demolition contractors that would have driven up prices. The programme's key success came from an extensive community engagement programme. A separate contract was formed for each property and this allowed the programme to be tailored to each property owner's timing and site-specific needs.
- A temporary toilets-and-showers programme involved the provision of more than 450 units delivered to properties, linked to cleaning and disposal services. This initiative supported people who chose to remain on their property, assisting them to maintain an adequate level of hygiene. The programme was originally designed to run for 12 months, but was extended twice to meet the needs of those who were still rebuilding, and the option was given for people to continue the contract beyond the end date of March 2011.
- The RAS was established based on the initiative of the building industry to provide rebuilding advice and to assist people in navigating through the complex rebuilding process. Two teams of roving RAS advisers were employed who could provide face-to-face advice, or help over the phone, often meeting people at their homes. Other forms of support included help for land surveys, particularly to assess future bushfire risk. The service proved popular, with support provided to nearly 1,000 households through 4,300 consultations, as of June 2011.
- In addition, a range of facilities was provided—communal laundry rooms, kitchens and dining halls, storage areas, pet holding areas etc.—often through donations from companies and non-profit agencies. Many such agencies supported with block clean-up and maintenance activities.

Findings after three years

After actively supporting the post-bushfire recovery process, VBRRA ceased operations in June 2011, after almost two-and-a-half years, with the staff seconded from different government departments mostly returning to previous posts. A small body, the Fire Recovery Unit (FRU), was established to carry on the support to communities that were still recovering.

It was initially considered that, after two years, most of the housing reconstruction would be complete or nearing completion and that the role of the FRU would focus on referral, community capacity building, advocacy and monitoring.

However, a telephone survey conducted in 2012 revealed that while 80.6 per cent of 1,380 respondent households were in permanent accommodation, 19.4 per cent were still in temporary accommodation (see Table 2.1) (Fire Recovery Unit, 2012). Of the 19.4 per cent in temporary accommodation, about one-third were currently rebuilding, another one-third were intending to rebuild but had not yet started, and 1.5 per cent were converting a temporary structure. However, 28.3 per cent remained undecided more than two years after the bushfires (see Table 2.2).

This was the first extensive survey since the 2009 bushfires, indicating the housing outcomes for the bushfire-affected communities, and it highlighted the protracted process encountered in varying degrees by many of those rebuilding. The survey also highlighted the lengthy period of time living in temporary housing for some people, and it raises a question about the nature and use of temporary accommodation post-disaster. Rather than being purely a short-term solution, it appears that a significant number of people may have spent extensive periods in temporary accommodation, although the lack of prior housing surveys limits the conclusions that may be drawn.

TABLE 2.1 Percentage of housing status of post-bushfire households.

Status	%
Rebuilt on-site	43.6
Permanent off-site	37.0
Temporary on-site	10.4
Temporary off-site	9.0
TOTAL	**100**

Source: Fire Recovery Unit, 2012.

TABLE 2.2 Percentage of post-bushfire households in temporary accommodation by intention.

Intentions	%
Currently rebuilding	38.1
Converting temporary structure to permanent	1.5
Intend to rebuild, not yet started	32.1
Undecided/unknown	28.3
Total	**100**

Source: Fire Recovery Unit, 2012.

Lessons and recommendations

Further research is needed into the long-term outcomes for individuals and families affected by the 2009 bushfires relating to their choices and actions for remaining within a community or moving to another.

While helping people re-establish their lives, governmental and other agencies should be careful not to rush, so that people have the time to consider carefully their options and choose what they consider best for themselves. Individuals and families should not feel forced into rebuilding, either through perceived pressure or the assistance that often focuses on immediate rebuilding rather than other long-term housing options. This necessitates that the role of government and non-governmental organisations (NGOs) should be to assist people in realistically evaluating the range of available options, instead of advocating any specific and defined outcome.

Without sufficient evidence to guide policy either way, the role of government should be to ensure that there is clear information about the possible options so that households can make informed decisions about what would be most suitable for them. It is highly important to understand that such decisions may change over time and so continuation of this support in the months and years following the disaster might be necessary. Where additional government or other donated assistance is available, there might be an opportunity to provide tailored local accommodation options and so reduce some of the pressure to rebuild before informed decisions are made.

The following lessons point to directions for agencies, governmental or otherwise, that seek to assist post-disaster recovery in Australia and similar contexts elsewhere:

- Getting the timing of activities right is important and will rely on comprehensive prior planning. For example, providing support for re-surveying of boundaries prior to people replacing fences may seem like a good idea, but most people, concerned about livestock welfare and public liabilities, replace fencing as one of the first rebuilding activities. Subsequent disputes may arise on property boundaries, adding to stress and costs. Within the planned timeline of activities, advisory services should be made available early, when people are making critical decisions.
- Government should invest in planning for post-disaster (housing) rebuilding prior to future events, so that when a disaster occurs there is strong coordination of the wide range of supports, services and policies required to assist affected households rebuild or find other long-term housing options.
- If the process for provision of financial assistance, whether through government or appeal funds, is too bureaucratic, long timelines can result while the building schedules are tight. This causes a lack of focus on long-term housing outcomes, unless targeted specifically to individual household needs. How such targeting can be done effectively is worthy of more research.

- Better interim housing options are required to protect the household economy of those rebuilding and to provide them a wider array of choices. There were people who spent a significant amount of money on making a shed inhabitable, for which they will never be able to obtain a building permit. This points to the importance of advice and support for interim solutions. For example, a community in the bushfire-affected town of Strathewen organised mobile 'holiday cottages' that maintained some resale value and were more comfortable and accessible than a caravan.
- Options and advice on lower-cost intermediate housing options that can develop into long-term housing can prove valuable after a disaster. For example, after Hurricane Katrina in the USA, the 'Katrina Cottages' were developed in a variety of styles as temporary housing that could be incrementally extended over time. This approach attempted to blur the distinction between temporary and permanent housing, providing a reasonable quality of life over the short term, while leading to longer-term habitation. The 'starter' cottage was a 'core house', providing basic household safety and accommodation needs, which could later be incorporated as a safe unit into a larger house. A core housing unit that is sufficient for immediate needs is economical and can be further developed as time and money become available is a concept worth exploring in the Australian context.

Conclusion

The complex nature of post-disaster recovery and rebuilding means that there are no easy solutions to these difficult housing issues. However, there is a clear need for a suite of initiatives that enable communities and individuals with a wide range of needs to achieve a long-term housing outcome that is in their best interest. Rebuilding of housing is not a race: outcomes should focus on individuals and families being able to choose the right options for long-term housing and community suitability, rather than focusing solely on rebuilding destroyed houses. The lengthy period of time that many people may need to live in temporary accommodation requires rethinking the use of marginal on-site accommodation in caravans in (for example) temporary villages. It points to the necessity for making available different options that allow more choice and also provide a reasonable standard of living for a long period of time.

Within the array of support from the government and NGOs, there exists an opportunity for built environment and design professionals to avoid focusing in isolation on the final house as a product. Such products, though well designed, when separated from context may be unaffordable or not appropriate in the context of the difficult post-disaster dilemmas that disaster-affected people experience. Instead, housing options that blur the distinction between temporary and permanent should be explored. Such housing could be quick to build, offer an acceptable quality of life, be affordable for most people (thereby reducing dependency on extensive government or donor support) and be flexible for adaptation to future use. Community engagement would need to be an intrinsic element in such rebuilding initiatives, whereby housing options are matched to the diverse needs of community members, yet contribute to the overall well-being of the larger community. Having lived the experience of a disaster and then facing the immediate need for some form of accommodation, often with limited financial resources, people do need some form of support from the government and other agencies. But they have to be partners in the process with the mutual aim of achieving safe, long-term housing outcomes.

Acknowledgement

Many of the findings of this chapter have been reported by the authors in a different form in a paper published in *Open House International*, 39(3). This chapter has been adapted from that paper in a significantly revised and updated form.

References

Bonnanno, George A. *et al.* (2010) 'Weighing the Costs of Disaster: Consequences, Risks, and Resilience in Individuals, Families, and Communities'. *Psychological Science in the Public Interest*, 11(1), pp. 1–49.

Charlesworth, Esther and Ahmed, Iftekhar (2015) *Sustainable Housing Reconstruction: Designing Resilient Housing after Natural Disasters*. London: Routledge.

Evans-Cowley, Jennifer and Kitchen, Joseph (2011) 'Planning for a Temporary-to-Permanent Housing Solution in Post-Katrina Mississippi: The Story of the Mississippi Cottage'. *International Journal of Mass Emergencies and Disasters*, 29(2), pp. 95–131.

Fire Recovery Unit (2012) *Fire Recovery Housing Survey*. Available at: www.rdv.vic. gov.au/_data/assets/pdf_file/0009/196137/Fire-Recovery-Unit-Housing-Survey-2012-v1.0.pdf (accessed 31 March 2014).

Olshansky, Robert B. (2006) 'Planning after Hurricane Katrina'. *Journal of the American Planning Association*, 72(2), pp. 147–153.

Olshansky, Robert B. *et al.* (2008) 'Longer View: Planning for the Rebuilding of New Orleans'. *Journal of the American Planning Association*, 74(3), pp. 273–287.

Paul, Bimal and Che, Deborah (2011) 'Opportunities and Challenges in Rebuilding Tornado-Impacted Greensburg, Kansas as "stronger, better, and greener"'. *GeoJournal*, 76(1), pp. 93–108.

River, Jeneden (2013) Personal communication.

Sapat, Alka *et al.* (2011) 'Policy Learning and Policy Change: Katrina, Ike and Post-Disaster Housing'. *International Journal of Mass Emergencies and Disasters*, 29(1), pp. 26–56.

VBRRA (2011) 'Legacy Report'. Victorian Bushfire Reconstruction and Recovery Authority. Available at: http://trove.nla.gov.au/work/159343996?selectedversion=NBD 51051068 (accessed 31 March 2014).

Zhang, Yang and Peacock, Walter Gillis (2010) 'Planning for Housing Recovery? Lessons Learned from Hurricane Andrew'. *Journal of the American Planning Association*, 76(1), pp. 5–24.

3

A PARTNERSHIP-BASED COMMUNITY ENGAGEMENT APPROACH TO RECOVERY OF FLOOD-AFFECTED COMMUNITIES IN BANGLADESH

Iftekhar Ahmed

Introduction

Unusually heavy monsoon rainfall in 2004 caused a massive and extensive riverine flood in Bangladesh, submerging more than 60 per cent of this country of predominantly low-lying floodplains comprising one of the world's largest deltas (BBC News, 2004). All the three main rivers—Padma (Ganges), Meghna and Jamuna (Brahmaputra)—and their numerous tributaries spilled over their banks, causing great loss and misery in the impoverished country. Nearly 750 deaths were recorded and the economic loss was more than US$2.2 billion. About 600,000 houses were totally destroyed and more than 3 million houses were damaged to varying degrees (DER Sub-Group, 2004; Planning Commission, 2008). This was a slow-onset disaster, with inundation lasting for more than two months. Communities were displaced from their settlements and sought refuge in temporary roadside shelters, relief centres (such as schools) or other makeshift arrangements. The flood destroyed people's physical assets (house, workplace, livestock, etc.) and also affected their livelihoods and economic conditions because of its prolonged duration. Hence the disaster had a two-pronged impact, necessitating a two-pronged physical and economic approach to recovery.

The United Nations Development Programme (UNDP) launched a 'Flash Appeal' to international donors for US$210 million for the colossal flood recovery, but only about 23 per cent of this amount was received even three months after the flood ended (Lockwood and Barnham, 2004). The Disaster Management and Crisis Prevention Team (DMCPT), under UNDP's Disaster Response and Recovery Facility (DRRF), was nevertheless established to work with the limited resources to benefit as many flood-affected people as possible. DMCPT brought in managers and specialists to design, manage and implement the DRRF over a one-year period in 2004–2005. The programme was implemented within the wider

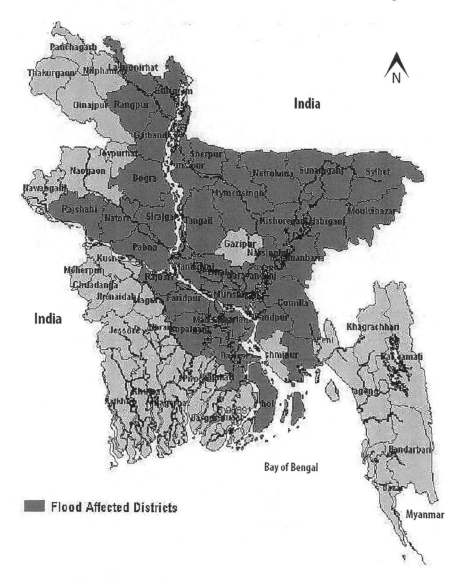

FIGURE 3.1 Map of Bangladesh showing extent of the 2004 flood.
Source: Adapted from ReliefWeb, 2004.

financial, legal and institutional framework of UNDP and supported in these areas by UNDP's technical and administrative personnel. The author was the Shelter Specialist for DMCPT, and this chapter has been derived from primary and secondary programme documents, field studies and other related information.

This case study is based in one of the most disaster-prone, climate-change-threatened and impoverished countries of the world (IRIN, 2009). The extensive disaster impacts in Bangladesh necessitate institutional and humanitarian support and there is a wide range of agencies active in the field, particularly local and international NGOs (non-governmental organisations) (Begum *et al.*, 2004). Thus, the approach to community engagement in the programme involved building partnerships between stakeholders at different levels. Because of the special post-flood emergency and humanitarian crisis circumstances, to expedite the programme, it was allowed to be implemented under the Direct Execution (DEX) modality where UNDP had the full responsibility and provision for engaging partner NGOs (PNGOs) that had a direct connection to the target communities. In its basic form, it was a tripartite partnership between UNDP, PNGOs and communities. However, the stakeholder network was much wider, also involving the national government and local government authorities at different administrative tiers, particularly in local level needs assessments, consultations, beneficiary selection, etc.

The approach to community engagement here was underpinned by the tripartite stakeholder partnerships, where PNGOs were the main conduit between the community and UNDP. There were also direct community engagement opportunities for UNDP through its monitoring, technical supervision and capacity building activities. Hence, all three key stakeholder groups had opportunities to interact with each other, offering the possibility of a transparent community engagement process.

While this chapter presents a case study of a programme implemented more than a decade ago, a wider set of impacts became apparent over the following years, particularly under the Shelter Cluster initiatives coordinated by UNDP and the International Federation of Red Cross and Red Crescent Societies (IFRC). Recent explorations indicate significant improvement in the economic and physical conditions of the beneficiary communities over the last decade, perhaps catalysed to an extent by the DRRF.

Key operational aspects

The programme structure and stakeholder partnerships were two key operational aspects in the DRRF. The programme structure was based on the different short-, medium- and long-term needs of flood-affected communities. A programme such as this, involving large numbers of beneficiaries and implementing agencies, relies on partnerships between different stakeholders, as discussed below.

Programme structure

The DRRF was implemented in two main phases: first, a three-month flood 'response', including emergency and relief operations during the later stages of the flood and its immediate aftermath, followed by a nine-month 'recovery' that included a 'cash-for-work' component for economic recovery and a 'family shelter' component for physical recovery. There was a degree of flexibility; for example, some response projects were provided additional funding and extensions because of their good performance, and they continued even after the recovery projects had started.

Asset transfer was a key aspect of all the programme components. Assets considered vital and necessary at the different stages of emergency and recovery were provided to the most severely affected communities. At the response stage, relief items were distributed based on needs assessment consultations with affected communities. In the recovery stage, direct cash and houses were provided. The asset transfer approach aims to go beyond addressing consumption and emergency needs by providing building blocks for future community well-being and resilience. It involves flexibility in accepting that assets provided may not be used by beneficiaries as intended by the implementing agency, but nonetheless benefit them in other ways. In subsistence-level communities that are vulnerable to begin with and which become even more vulnerable after a disaster, it may appear that most asset transfers tend to provide benefits in some way, although these may involve adaptation or alternate use of the asset, or outright sale. However, community engagement provides the opportunity to provide assets that would more directly meet the needs of the beneficiaries.

Partnership arrangements

For the DRRF to be implemented within its limited budget, a selection process had to be undertaken to engage the most suitable PNGOs out of the many applications from Bangladesh's large NGO community. Vetted by DMCPT members, the selection involved assessing the applicant NGO's relevant experience and implementation capacity. Based on further assessment of detailed project proposals, a final group of PNGOs were selected. For the recovery component, a key underlying element guiding the selection process was that the PNGO should have existing strong relations and engagements with the target flood-affected communities. Partnerships were central to the DRRF as a vehicle for community engagement, particularly partnerships with NGOs that were often better placed for community-based projects than UNDP or the government. Such partnerships are relatively easy to establish in Bangladesh because NGOs are highly active throughout the country (see for example, Begum et al., 2004). Thus implementation arrangements with PNGOs were a key operational aspect.

For the recovery programme, partnerships were formed with five selected PNGOs: Bangladesh Rural Advancement Committee (BRAC); Gono Shastha Kendra (GSK/Public Health Centre); Islamic Relief Bangladesh (IRB); National Development Programme (NDP); and Oxfam Bangladesh. The role of UNDP was to manage the programme funding and coordinate the implementation of individual projects of the PNGOs, and to provide technical, supervisory and monitoring support. The individual PNGO projects were planned according to organisational capacity, the nature of project locations and avoidance of overlapping locations, and the extent of existing community engagement of the PNGO. Midway through the programme, funding became available to engage three more PNGOs with a smaller scope of work. Here it was possible for them to build on lessons from the projects the other PNGOs had implemented earlier.

Selecting these agencies is often fraught with difficulties in a context where a multitude of diverse NGOs is active; in 2000 there were more than 1,100 registered NGOs in Bangladesh (ADAB, 2000). Factors such as political affiliations and rivalries, conflicting agendas and varying levels of community engagement create challenges in selecting and coordinating a group of very different PNGOs. Would a large, well-established, experienced NGO be a more reliable partner than a small NGO that is nonetheless very much locally based and has a strong link with local communities? Most NGOs tend to focus on socio-economic development, so would they be able to deliver physical outputs such as housing and infrastructure? Questions such as these could not be resolved clearly at the selection stage, but they were often answered by lessons learned at subsequent stages.

Phase 1: Response

Emergency response actions were taken following the flood to meet the basic needs of the affected people. Initial relief assistance was provided for emergency survival needs: water and sanitation, healthcare, clothing and essential household items.

Because of the prolonged flood and the time and logistics involved, some of the later response activities could be considered early recovery assistance. This included provision of crop seeds, house repair, and educational support by payment of exam fees for students. Timing played a crucial role in determining such needs in the different post-flood stages.

Table 3.1 shows an outline of the DRRF response component, indicating the partnerships with a large number of different PNGOs for delivery of different types of relief assistance. The PNGOs' direct engagement with the beneficiary

TABLE 3.1 Summary of post-flood response by PNGOs.

PNGO	Time (months)	Type of assistance	Target (Households)
ActionAid	4	Medicine; House repair; Tubewell and latrine repair; Seeds (home gardens and paddy); Education fees (SSC and HSC candidates)	15,000
ADRA	1	Food	30,000
BD Scouts	4	ORS; Seeds	1,500,000; 30,000
BRAC	2	ORS; IV Saline; Medicines; Alum; Life jackets; Matches; Candles; Carbolic soap; Mosquito nets	100,000

PNGO	Time (months)	Type of assistance	Target (Households)
CARE	6	Paddy and veg. seeds	130,000
Christian Aid	1	ORS; Alum; Food	500,000
Concern	0.5	ORS; Alum; Tubewell repair and cleaning	31,500
GSK	3	Medical aid; Cattle feed; House repair; Seeds	50,000
Handicap International	6	House repair/reconstruction; Ground raising of social institutions; Latrines and tubewell installation; Seeds; Disabled support	4,000; other units
ICCDDR, B	3	Medicines; ORS; Beds; Buckets; Bowls; Fans; Cleaning items to hospitals	91,500
Islamic Relief	6	Paddy and veg. seeds; House repair; Cash grants	46,000
NDP	1	Soap; Napkins; ORS; Food	30,000
Oxfam	2	Water purification tablets; Latrines; Tubewells; ORS; Soap; Mosquito nets; Alum; Fodder; Clothes; Cotton towels; Veg and crop seeds	28,800
Save the Children (UK)	3.5	Safe water; Sanitation; Seeds; House repair	85,000
Save the Children (USA)	3	Safe water; Sanitation; Agricultural support	48,000

communities allowed local consultations to identify suitable relief items and target them where needed most. UNDP coordinated the PNGOs to avoid overlapping of target areas and to distribute a range of items as required in different areas and communities. Through the different response activities, nearly 3 million flood-affected people benefited.

Phase 2: Recovery

The DRRF's nine-month post-flood recovery phase followed the three-month response, although some 'early recovery' response activities continued, as mentioned above. As highlighted by UNISDR (2009): "Recovery programmes, coupled with the heightened public awareness and engagement after a disaster, afford a valuable opportunity to develop and implement disaster risk reduction measures and to apply the 'build back better' principle" (p. 23). This was reflected in the DRRF recovery programme, which incorporated disaster risk reduction (DRR) in the process of restoration and improvement of flood-affected community facilities, living conditions and livelihoods. DRR was a key feature of the recovery programme, with the objective that beneficiaries would be less vulnerable to the next floods due to raised homesteads, repaired dykes, flood shelters and flood-resistant housing (see UNDP, 2005), contributing to their long-term resilience.

For the recovery programme, through a range of needs assessments and consultations at the local and institutional levels, restoring and improving the local economy and extensively devastated housing were prioritised and an asset transfer approach was followed; that is, providing the economic and physical assets necessary after the flood to enable recovery. Economic and physical recovery was supported by the 'cash-for-work' and 'family shelter' programme components respectively.

Cash-for-work

The post-flood economic recovery focused on livelihoods through cash-for-work (CFW) activities targeted predominantly at women in order to address their specific vulnerability in a male-dominated society; more than 87,000 or about 80 per cent of the CFW beneficiaries were women. Additionally, other extremely poor people—including men, the elderly and disabled people—benefited. More than 250,000 people received income-earning opportunities, ranging from three weeks' to three months' duration in 25 severely flood-affected districts. The CFW initiative was targeted at vulnerable and marginal people who would face barriers in finding work in the open labour market. It aimed to reduce economic hardship and prevent people from sliding further into poverty. As intended, it prevented people from 'distress selling' their assets for want of cash for basic needs.

Beneficiaries were selected by the PNGOs in consultation with local government authorities to be workers in earthworks, such as excavating or enlarging drainage canals, digging ponds and raising dykes, roads, markets, schools and homesteads. This significantly improved the community infrastructure and helped prevent future flooding and waterlogging. The largely women workers were paid daily wages at the local market rate, allowing them to bring income to their households and to earn a fair wage, not be exploited because of their vulnerability, particularly after a disaster. In a context where women have limited access to, and control of, money, the cash transfer contributed to the empowerment of the female beneficiaries. Digging the canals and raising community facilities reduced the risk from future floods, and at the same time injected much-needed funds into the flood-damaged local economy. Thus the economic recovery also enabled disaster risk reduction.

Rural homesteads are typically built on raised earthen mounds to protect them from the regular floods in this low-lying country. However, the 2004 flood was unusually massive and breached many homesteads. Therefore, through CFW, the elevation of low-lying homesteads was raised by digging and piling soil. This was complementary to the family shelter rebuilding, which was often built on homesteads raised through CFW. Thus, in combination, the CFW and family shelter components contributed to disaster risk reduction.

FIGURE 3.2 (a) Digging a drainage canal through CFW; (b) An improved house built on a homestead raised through CFW.

Family shelter

It was a strategic decision in DMCPT to provide family shelters that were an improved version of the typical rural house, partly by ensuring that good-quality materials are used. Brick and other durable materials would have been far more expensive than the local materials—timber, bamboo, etc.—and fewer houses could have been provided. This was the largest programme of its kind, building more than 16,000 houses, yet this number was low compared to the 4 million houses damaged by the flood. Thus, it was decided to stretch the funds to support as many people as possible in a context of great need.

Each family shelter was specified to be no less than 20 square metres, in line with local standards (GoB, 2008). Published in detail as an illustrated handbook (Ahmed, 2005), the key improvements of the family shelter over typical rural houses included:

- Stabilised plinth: Typical earthen plinths were raised and stabilised with a capping of soil-cement for better durability and resistance to floods.
- Concrete footings: These were used to prevent decay of timber/bamboo posts by separating them from the ground. Some of the PNGOs chose to use full concrete posts instead of the combined, and less expensive, timber/bamboo post and concrete footing version by investing additional funds from their own sources. This was done to serve as a sturdier option for windy areas.
- Bamboo treatment: Bamboo for structural elements and wall panels was treated against insect attack. A simple technique of injecting odorous oil (kerosene, diesel, etc.) into bamboo to repel insects was used, prior to which bamboo was soaked in water for at least a week to dilute its sap.
- Rainwater gutter: A simple system of split PVC pipes was used to make roof gutters for draining and collecting rainwater, useful in areas facing arsenic contamination of sub-soil water (Hossain, 2006). Gutters also prevented damage of plinths from water dripping from roof eaves.
- Corrugated iron roofing sheet: To prevent lift-off and damage by wind, a minimum thickness of 0.35 millimetres was specified for the corrugated iron (CI) roofing sheets.

The PNGOs were provided training by DMCPT, both for the CFW and the family shelter components. The shelter training included hands-on practical sessions on plinth stabilisation and bamboo treatment. The PNGO staff then trained local construction workers and beneficiary household members to build the houses. DMCPT staff made regular field visits for technical supervision and monitoring. However, constructing more than 16,000 houses in 23 districts meant that only a sample of the housing could be inspected. The monitoring and supervision were thus organised according to each PNGO, so that any lessons from particular sites could be shared across other project locations under the same PNGO.

FIGURE 3.3 The UNDP houses, such as the one on the right, were an improved version of typical rural houses such as the other houses in the photo.

While maintaining the basic family shelter design guidelines, there was flexibility in making variations (without compromising quality), for example, using materials other than bamboo mat walls as specified; walls of various types of materials were built according to local availability; concrete posts instead of concrete footings were used in some cases. The extensive community engagement process allowed decisions to be made on what the locally appropriate options were.

Building a large number of houses brought work and income opportunities for a wide range of local businesses, construction and related workers, building materials suppliers, etc. Even local transport workers earned income from carrying building materials and products. In this way by bringing much-needed funds to the local economy, the physical recovery also supported economic recovery.

In terms of physical recovery, beyond the family shelter, flood–damaged infrastructure (roads, school grounds, earthen mounds/flood shelters, dykes, market places, etc.) in 24 severely flood-affected districts were repaired and raised above flood levels with essential turfing and compaction. A number of community flood shelters were built, which normally function as schools. Thus, the recovery process incorporated DRR into the repair and reconstruction process so that the communities' physical assets were better protected from future floods.

Opportunities and challenges

The recovery programme focused on cost-effectiveness, that is, cost was minimised without compromising quality. Given the limited resources available to assist a huge number of flood-affected communities, the opportunity was there to make decisions to channel CFW funds through women in order to ensure family well-being and to provide good-quality bamboo-and-corrugated-iron family shelters instead of brick to reach more beneficiaries. The programme also took the opportunity of utilising local materials so that the houses could be built in a rural context without the need for highly skilled workers, who are typically male, hence providing scope for women's involvement. Very little training was required for both the CFW and the family shelter components because they were derived from existing local practices. Importantly, the recovery programme was utilised as an opportunity to improve the economic and physical conditions so that communities could become more resilient to future floods.

Despite such achievements, implementing such a large-scale programme is bound to encounter challenges in a developing country like Bangladesh. Some of the key challenges were:

- Community engagement: The PNGOs were the conduit for community engagement given their relevant experience in working directly with communities, but results varied because of the differing levels of engagement of the different PNGOs. While the smaller and local PNGOs seemed to have a closer engagement, the larger PNGOs had more resources and were able to deploy more staff for longer durations in the field.
- Implementation process: While some PNGOs provided contractor-built houses, others followed an owner-driven process by providing funds and sometimes building components, and beneficiaries arranged the construction backed by technical support. Both approaches had advantages and disadvantages, but the latter resulted in better beneficiary satisfaction, although often requiring more time. For the CFW, sometimes there was local resentment that more women than men were prioritised. In a strongly patriarchal society like Bangladesh, such issues were challenging and required extended negotiations with the community.
- Beneficiary selection: Nepotism and corruption are endemic in Bangladesh; the local government authorities assisted in the selection of beneficiaries, but there were allegations of unfair selection. Also PNGOs prioritised their existing members, who were not always the neediest. They were receiving some form of support already from the PNGO, unlike others who did not receive any form of assistance, especially landless people in informal settlements. PNGOs also preferred to provide support in areas that were easier to access and avoided hazardous locations, although these were often the places where help was most needed.
- Project planning and logistics: Large-scale production and processing of building components and materials, and transporting, storing and distributing

them presented a wide range of logistical challenges, particularly for the smaller PNGOs. Adequate staffing and training were also challenges, particularly for the PNGOs with less experience of recovery projects.

• Quality control: The programme had a strong emphasis on ensuring good quality of materials and construction. Caution, particularly with concrete and CI sheets, was essential, sometimes leading to disputes. Corruption in the construction field is endemic in Bangladesh, requiring strong vigilance and monitoring. Supervision of such a large-scale programme was a key challenge, and it was exceedingly difficult to monitor construction quality in so many sites spread across the country. Relying on PNGO staff, although they were provided with training, meant that it was often hard to determine whether capacity building on improved house construction had been achieved.

Long-term impact

Some of the PNGOs have continued using the ideas developed in the programme. For example, IRB has used concrete footings and treated bamboo as used in the UNDP project in a Rohingya refugee camp in southern Bangladesh. After Cyclone Sidr in 2007, the Shelter Cluster, coordinated by UNDP and IFRC, produced training manuals in which some of the technical innovations of the DRRF were included. These reached a wide section of the NGO community involved in the post-disaster shelter provision.

However, brick and CI sheets have now become more widespread with increasing urbanisation and have also become more affordable. In consequence, bamboo and other natural materials have greatly declined in use. People's aspirations have changed and they now expect houses built of "modern" materials in recovery programmes, instead of the modest houses provided in the DRRF. Even agencies prefer such materials as the most resilient form of housing and many prominent agencies provide brick houses after disasters. This is partly because the agencies do not want to provide houses that would be destroyed in the next disaster.

More than ten years later, field visits to some of the project areas generally indicated overall economic improvement, particularly in the areas near major cities. Houses were much bigger and most had CI sheet walls and roofing, and brick walls in some cases. By this time, the UNDP houses had been dismantled and replaced by bigger houses. Some of the materials, such as bamboo, had decayed over the years and materials that could be re-used were utilised or stored for the future. Particularly, because the CI roofing sheets were thick and of good quality, they were re-used for the new house roofs and, in most cases, were the only building components that were used as intended. Therefore, specifying and enforcing a minimum grade of CI sheet and the transfer of this particular asset to flood-affected households resulted in long-term beneficial outcomes.

A house in a rural and low-income community in Bangladesh usually follows an incremental process of extending and upgrading; it is seldom a completed product.

FIGURE 3.4 This much larger CI sheet house replaced the UNDP house that stood on this site.

Thus the UNDP family shelter served as the basis for household investments and improvement initiatives over time following that incremental process, so that after a decade, the flood-affected communities were in a more stable and improved situation.

The DRRF family shelters were provided in the aftermath of great devastation caused by an unusually severe flood, which left people in dire need of asset replacement. Together with the CFW, it helped stabilise a disrupted community, so that it could gradually move forward towards improved prosperity and resilience. The improved housing and economic conditions observed recently were built on, even if to a limited extent, with the support provided through the recovery programme.

Conclusion

For a large-scale, time-bound programme such as this, community engagement has to be understood differently from a small-scale, grassroots, community-based programme. For direct community engagement in a large programme, more staff and resources are required and it becomes impractical for any one agency, even for a significant agency such as UNDP, to operate alone. This creates the need for a partnership strategy involving multiple agencies, and community engagement becomes more feasible when the partners have existing links and experience of working with the communities concerned. Large agencies such as UNDP are seldom in a position to engage directly or extensively with local communities, partly because of high personnel and other costs in line with international standards. The DRRF programme enabled UNDP to work with local communities via the partnerships with PNGOs. However, this strategy requires intensive monitoring, technical supervision, quality control and, importantly in a context such as Bangladesh, dealing with corruption. Where corruption is pervasive and endemic, avoiding it altogether might not be possible. Yet when programmes such as the DRRF are implemented by agencies with high global standards, efforts are made to limit corruption as much as possible. In the process, disputes escalate and organisational relationships become tainted, but the eventual outcome is that benefits increase for programme's beneficiaries. A small example is the effort in the DRRF programme to ensure that good-quality CI roofing sheets were provided, despite attempts to compromise the standards, and the benefit of this was apparent ten years later.

Assisting 3 million people and building more than 16,000 houses is a large programme by anyone's standards. However, the 2004 flood affected 35 million people—a quarter of Bangladesh's population—and 4 million houses were damaged or destroyed. Despite its scale, the DRRF programme could cater only to a fraction of the vast number of flood victims, the majority of whom had to arrange their own recovery, often in *ad hoc,* makeshift and inadequate ways, which left them vulnerable to future floods. With climate change resulting in an increasing extent and frequency of floods, more and more people will be affected in the future, and agencies will not be able to assist all, or even most, of them; the onus of recovery will have to be on communities themselves. Therefore, future approaches to community engagement need to emphasise disaster risk reduction through building community resilience and capacity for effective self-recovery. Institutional policy and practice, including those of prominent international agencies such as UNDP, need to move beyond an asset transfer approach to enabling communities to take control of their own recovery process.

Acknowledgement

This chapter is drawn from work undertaken by the author as a specialist member of the Disaster Management and Crisis Prevention Team (DMCPT), UNDP, Bangladesh.

References

ADAB (2000) *Directory of NGOs.* Dhaka: ADAB (Association of Development Agencies in Bangladesh).

Ahmed, Iftekhar (2005) *Handbook on Design and Construction of Housing for Flood-Prone Rural Areas of Bangladesh.* Bangkok: Asian Disaster Preparedness Center.

BBC News (2004) 'Bangladesh Appeals for Flood Aid'. Available at: http://news.bbc. co.uk/2/hi/south_asia/3530532.stm (accessed 1 February 2016).

Begum, Syeda Feroza, Zaman, Sawlat Hilmi and Khan, Shahin Mohammad (2004) 'Role of NGOs in Rural Poverty Eradication: A Bangladesh Observation'. *BRAC University Journal*, 1(1), pp. 13–22.

DER Sub-Group (2004) *Monsoon Floods 2004: Post-Flood Needs Assessment Summary Report.* Dhaka: DER (Disaster and Emergency Response) Sub-Group.

GoB (2008) *Construction of Sidr Core Shelters, Minimum Standards.* Dhaka: GoB (Government of Bangladesh).

Hossain, Mohammed Faruque (2006) 'Arsenic Contamination in Bangladesh: An Overview'. *Agriculture, Ecosystems and Environment*, 113(1–4), pp. 6791–6800.

IRIN (2009) 'Twelve Countries on Climate Change Hit List'. Integrated Regional Information Networks. Available at: www.irinnews.org/report/85179/global-twelve-countries-on-climate-change-hit-list (accessed 1 February 2016).

Lockwood, Harold and Barnham, Jane (2004) *UN Flash Appeal Bangladesh Floods 2004: Independent Review Mission Report.* Available at: www.lcgbangladesh.org/flood2k4/docs/UN%20Flash%20Appeal%20Review%20-%20Final%20Report_Nov%2004.pdf (accessed 15 April 2016).

Planning Commission (2008) *National Strategy for Accelerated Poverty Reduction II.* Dhaka: Planning Commission Government of People's Republic of Bangladesh, General Economics Division.

UNDP (2005) *Fix the Risk: UNDP's Flood Relief and Recovery Program* (video). Dhaka: UNDP (United Nations Development Programme).

UNISDR (2009) *Terminology on Disaster Risk Reduction.* Geneva: UNISDR (United Nations International Strategy for Disaster Reduction).

4

PUBLIC ENGAGEMENTS IN FORWARD-LOOKING RECOVERY EFFORTS FOLLOWING THE 2013 FLOODS IN HIGH RIVER AND CALGARY, CANADA

Eva Bogdan, Amber Bennett and Lilia Yumagulova

Introduction

The 2013 flood in the Province of Alberta displaced 100,000 people and became Canada's costliest natural disaster at the time (at CAD$6 billion). All levels of government responded by creating goals relating to transformative change and community resiliency. Natural disasters are not unpredictable acts of nature, but rather a consequence of how society is organised (Freudenburg *et al.*, 2009; Mileti, 1999; Perry, 2007; Tierney, 2012). As such, changes at the civic and institutional levels, rather than technical innovations, are required to alter aspects of society that increase vulnerability to risks.

Creating intentional changes at the societal level is a monumental challenge and even more so in the aftermath of a disaster amid upheaval in neighbourhoods, institutions and infrastructure. Yet, post-disaster chaos may be the most opportune time for re-ordering and re-structuring a community, a society and a nation. We ask, did public engagements on the 2013 flood promote *forward-looking* recovery rather than a bouncing back to 'normalcy' in ways that retain vulnerability to the same hazards? To what extent was this disaster used as an opportunity to shift public engagement strategies towards empowering citizens for facilitating change? The purpose of this chapter is to give voice to participants' experiences and interpretations of public engagement in these recovery efforts and their concerns over root causes and systemic issues related to flood disasters.

In Alberta, public involvement in natural resource management issues has tended to lag behind other provinces (Sinclair and Diduck, 2001). By examining community engagement in the Alberta context, we seek to contribute to the dearth of literature on the social dimensions of flood management in Alberta (Bogdan, n.d.) and disaster scholarship addressing recovery, community resiliency and place (Cox and Perry, 2011). We begin with a brief background of the 2013 Alberta flood. Next, we review the literature on disasters, recovery and resilience;

public engagement timing and format; and flood management in Alberta. A description of the Town of High River (ToHR) and City of Calgary (CoC) contexts and research methods follow. Next, using evidence from each of the two case studies, we discuss public engagement during the 2013–2015 recovery phase. We conclude by suggesting improvements for public engagement.

Background

In June 2013, southern Alberta experienced the most damaging flood in recent history. An intense 72-hour period of rain, measuring in excess of 300 mm, fell on the headwaters of Highwood River, Elbow River and Kananaskis River (Pomeroy *et al.*, 2016)—impacting rivers that flow through the City of Calgary and the Town of High River (see Figure 4.1). The impact of the flooding was unprecedented in Alberta and resulted in the first-ever declared state of provincial emergency (Auditor General, 2015). The flood damage was extensive: affecting 55,000 square kilometres, impacting more than 100,000 people in more than 30 communities,

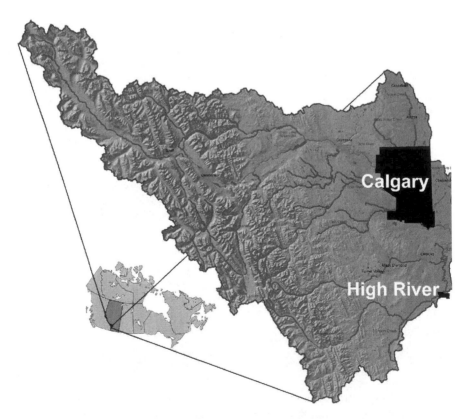

FIGURE 4.1 Map of Alberta showing the Bow River basin watershed and the municipalities of Calgary and High River.

Source: Aggregated and adapted from multiple images.

damaging 14,500 homes, resulting in five deaths, and costing CAD$6 billion in property and infrastructure damage (ibid.). While the 2013 flood was remarkable in its scale, floods in Alberta occur regularly. In fact, there have been 12 large floods in the past 135 years in the region where the two municipalities are located (ibid.), providing multiple opportunities to develop and refine disaster and emergency management plans.

The flood devastated the Town of High River, resulting in the most extensive damage per capita. High River has a population of 12,715 (Statistics Canada, 2014). Flowing through the town is the Highwood River, which has flooded ten times in the past century. The Highwood River flowed at more than 30 times greater than normal for the time of year. More than 13,000 residents were evacuated from the town and rural areas. Fifty-nine per cent of the land was inundated by water, 70 per cent of homes were moderately to severely damaged, and 79 of 83 town buildings experienced significant damage (ToHR, 2015a).

Calgary is located 37 kilometres north of High River and is Alberta's largest city, with a population of 1,195,194 (CoC, 2014a). Calgary incurred the most expensive damage during the flood. Approximately 6,000 homes sustained water damage from overland flooding, seepage or sewer backups and 4,000 businesses were impacted by the flooding, evacuations or power loss (Arthurs, 2015). Approximately 75,000 people had to leave at the height of the crisis.

Literature review

There is a body of literature on the importance of post-disaster public engagement, its advantages and challenges (see Vallance, 2011). There are fewer studies examining the different formats of public engagement post-disaster. Furthermore, there is a dearth of disaster scholarship addressing recovery, community resiliency and place (Cox and Perry, 2011). We aim to address these gaps in the literature by examining public engagements following the 2013 Alberta floods and whether they promoted 'forward-looking recovery' by creating deliberative spaces that allow for inclusion of, and collaboration among, key stakeholders to build resilient communities.

Disasters, recovery and resiliency

According to social constructivists, a natural disaster, while triggered by a hazard, is ultimately a social phenomenon (Perry, 2007). These disasters are "manifestations of failures in environmental governance and sustainability" (Tierney, 2012, p. 358); thus, solutions require changes in the social order and social activities (e.g. not building in floodways). In contrast, structural functionalist approaches focus on physical agents and hazards, and view disaster events as disrupting the social structure (Webb, 2007); thus, solutions involve controlling the hazard and managing the event (e.g. building dams). These differing philosophical perspectives shape how a disaster is framed and influence the choice of civic actions during recovery.

The disaster management process consists of four (non-linear) components: preparedness, response, recovery and mitigation/prevention. Recovery is the least understood component of the disaster management cycle (Smith and Wegner, 2007; Olshansky and Chang, 2009). A key tension during the recovery period is balancing swift action and the speed of rebuilding while carefully considering alternatives and involving a wide range of stakeholders through an appropriate level of engagement (Kim and Olshansky, 2014). The majority of resources are often spent on quick fixes to provide immediate assistance and protection.

However, a hasty return to 'normal' can reinforce vulnerabilities (Pelling, 2003), shortening and closing the post-disaster window for in-depth examination of the extent of damage, of root causes and of opportunities for future disaster risk reduction transformation at the community level. Echoing a social constructivist approach, Public Safety Canada (PSC) states that disaster risk reduction can be achieved through the following:

> Systematic efforts to analyze and manage the causal factors of disasters, including through the mitigation and prevention of exposure to hazards, decreasing vulnerability of individuals and society, strategic management of land and the environment, improved preparedness for disaster risks, coordinated response and planning and *forward looking* recovery measures.
>
> *(Public Safety Canada, 2011, p. 14; italics added)*

These efforts not only enable communities to recover "from recent disaster events, but also to build back better in order to help overcome past vulnerabilities" (ibid., p. 5).

Resilience is often discussed in the context of recovery and has been interpreted as *bouncing back* (Alexander, 2013), i.e. returning to and restoring the status quo ante. More recently, however, resilience has been interpreted by some scholars as *bouncing forward* to achieve positive gains (ibid.), which is more in line with *forward-looking* recovery measures. According to Public Safety Canada, "resilient capacity is built through a process of *empowering* citizens, responders, organizations, communities, governments, systems and society to *share the responsibility* to keep hazards from becoming disasters" (Public Safety Canada, 2011, p. 8; italics added). We investigate whether these notions of forward-looking recovery, empowerment, and shared responsibility were reflected in our participants' experiences.

Meaningful public engagement: Timing and format

Academic research emphasises the importance of community consultation and local participation throughout the recovery process (Kreimer, 1978; Rubin and Barbee, 1985; Berke *et al.*, 1993; Kim and Olshansky, 2014). Public Safety Canada echoes this: "Disaster mitigation is most effective when activities engage the community. Therefore, public awareness and education initiatives should be a priority" and should "promote a *culture* of mitigation . . . to affirm disaster risk reduction as a *way of life* for all Canadians" (Public Safety Canada, 2016, p. 3; italics added). Similarly, Shrubsole (2013) found that in Canada, improving flood management requires not technical innovations, but rather "a change in the *culture* and the *institutional arrangements* for flood management at all levels" (p. 117; italics added). However, engaging the public in meaningful[1] ways during post-disaster chaos (and, more broadly, in environmental issues which may lead to disasters) is complicated by capacity, diversity and place.

Effective public engagement post-disaster is seen as a challenge given the need to design engagement processes that would address decisions that are complex, replete with technical uncertainties, and characterised by perplexing value trade-offs (Dorcey and McDaniels, cited in Pearce, 2003). While authorities are often under immense pressure to start community engagement recovery processes immediately, research indicates that involving communities in complex decision-making processes immediately after a disaster event can be problematic, because those most impacted may have the least capacity (in terms of time, energy, finances and emotional reserves) to participate in constructive ongoing dialogue about long-term solutions (Ward *et al.*, 2008; Gordon, 2008; Spee, 2008). People need space and time to move through the different psychological phases of disaster response (see Figure 4.2). Thus the types of engagement formats and tools used, and when, should be given special consideration (Ward *et al.*, 2008).

Participatory processes can range from passive public involvement, in order to inform or "decide–announce–defend", through formats such as open houses, public hearings, written comments and registry systems. At the other end of the spectrum is active participatory engagement involving interaction and dialogue more common in formats such as workshops, site visits and working groups (Diduck and Mitchell, 2003; Sinclair *et al.*, 2008). For environmental land-use planning under non-disaster conditions, interactive and dialogical activities are critical during the

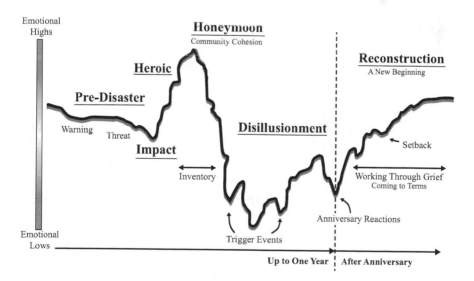

FIGURE 4.2 Phases of psychological reactions to disasters.

Source: Adapted from Zunin and Myers, as cited in DeWolfe, 2000 (permission from SAMHSA for reprinting).

normative (what *should* be done) and strategic (what *can* be done) stages, when learning is most likely to occur, rather than during the operational stage of planning (what *will* be done) which is when it typically occurs (Sinclair *et al.*, 2008).

Following a disaster "there may be benefits to delaying community input into long-term solutions until a later date when normal cognitive and social processes have been re-established" (Ward *et al.*, 2008, p. 18). However, delaying some aspects of the recovery may not be an option if critical infrastructure, necessary for the daily functioning of a community (such as roads and utilities) or to protect from damages from an imminent reoccurring disaster event, must be restored or built. Consequently, in such cases, decisions for designing and planning major infrastructure for the long-term (and to possibly build-back-better) cannot wait. When there are options to delay some aspects of recovery during the immediate stages following a disaster (but within the 24-month window of opportunity for change), public meetings that provide information and resources, as well as spaces for discussions, may be more important than asking for input on major decisions, as they help people to understand and make meaning of the event (Johnston *et al.*, 2012). Once communities move into reconstruction and participatory processes, engagement approaches must support them to 'come to terms' with what the future may hold. Moser (2012) argues that such engagement requires navigating terrain that is "political on the surface and personal—psychological, spiritual and cultural—deep underneath" (p. 4). Literature points to the critical need to be safe and supported during this stage of engagement so that people can notice their conflicts, fears, apathy and loyalties, and feel the pain of grief without being overwhelmed by the anxieties it generates. This typically involves two key practices: 1) creating a connection to others and feelings of support; and 2) creating a 'safe space'—be it a physical space or the atmosphere of how we communicate about the issues (Randall, 2009; Mnguni, 2010; Nicholsen, 2003; Weintrobe, 2013; Lertzman, 2014; Macy and Johnstone, 2012). Listening, dialogue and conversation are well suited to this sort of engagement. Weintrobe (2013) also argues that it is critically important that people feel supported by those responsible for leading and shaping the communities in which they live.

Spaces of support, safety and connection are also conducive to social learning.[2] Social learning is critical in complex natural resource management issues involving diverse interests. Through participation, an individual is able to reconcile their individual desires with those of others and in the process "learn to be a public as well as a private citizen" (Pateman cited in Hoverman *et al.*, 2011, p. 31). Studies have found that by exploring others' interests and seeking common ground, shared subjectivity can arise, transcending pursuit of individual interests. Participatory experiences increase the sense of integration into the wider community and society and can shift values towards the social and the environment (Craig, 1990; Hoverman *et al.*, 2011).

A place-based context is also an essential consideration in post-disaster recovery because communities are in closest proximity to hazards and thus can be change agents by reducing their vulnerability to hazards through land-use regulation and enforcement of building codes (Henstra and McBean, 2005). Public involvement and dialogue in parts of Canada (Manitoba and Ontario), as well as a collective learning approach in Europe (UK and Netherlands), have been deemed successful to reconcile diverse views and demands and to develop shared understandings of problems and potential solutions in flood management (Ashley *et al.*, 2012; Haque *et al.*, 2002; Hayward *et al.*, 2007; McCarthy *et al.*, 2011; van Herk *et al.*, 2011). In Alberta, public involvement in natural resource management issues in general has tended to lag behind other provinces when assessing for early involvement, learning opportunities, access to registry, adequacy of public notice, consideration of need and alternatives for a given project, and engaging diverse rather than just directly affected publics (Sinclair and Diduck, 2001). We examine the degree to which our participants felt public engagement in Alberta empowered citizens and facilitated change following the 2013 flood.

Flood management in Alberta

Historically, the Government of Alberta (GoA) has failed to pass laws to limit development in floodplains (Auditor General, 2015). Flood mapping identified 60 cities and towns across Alberta with an inundation hazard (IBI, 2015) but only 20 of those were designated as flood hazard areas (Auditor General, 2015). For example, an area of High River was identified in 1992 through the provincial mapping programme as flood-prone, but was not designated as such; it was later developed and sustained heavy damage in the 2013 flood. The lack of designation often reflects both lack of community support and GoA's reluctance to impose designation, leading to an inconsistent approach to managing development in flood hazard areas (ibid.) and, ultimately, resulting in a majority of citizens paying higher taxes and insurance premiums at the municipal, provincial and/or federal levels for flood damage to property and buyouts impacting only a minority of residents in the affected locations.

Following the 2013 flood, extensive structural measures (e.g. berms, dikes) and non-structural measures (e.g. updated emergency plans) were implemented in High River and Calgary, as well as the removal of two neighbourhoods in High River. GoA also enacted the *Flood Recovery and Reconstruction Act, Bill 27* (December 2013; see Alberta Municipal Affairs, 2015) to amend the *Municipal Government Act*, thereby restricting development in floodways to ensure that rebuilding occurs in ways that limit future flood risk. To date, the regulations have not been promulgated, therefore "municipalities still retain building decisions on their floodplains" (Alberta Municipal Affairs, 2015, p. 13).

GoA also developed a new Provincial Recovery Framework to promote a co-ordinated provincial approach to flood recovery, mitigate the risk of future floods and serve as a template for recovery efforts after future disasters (Government of Alberta, 2013a). The Framework consists of four essential elements of recovery—environment, social, infrastructure and economic—and describes how GoA will support local communities in their recovery efforts during the various stages. In this model, the local community is identified as the lead. While numerous public engagement events were held at the municipal and provincial levels,[3] our analysis

shows that the degree to which community members felt engagement was mean-
ingfully varied in the two case studies.

Methods

In this chapter, we examine public engagement during recovery from the 2013
Alberta flood through two different case studies. We chose Calgary and High
River because these municipalities sustained the most damage and thus provide
rich comparative potential: they are close to each other (both in southern Alberta,
where flooding is most frequent), yet their approaches to disaster recovery differed,
as have the outcomes.

Both case studies draw on in-depth, semi-structured qualitative interviews ranging from one to four hours, as well as researchers' observations through attendance at
community engagement events and staff presentations, analysis of media and historical and current documents and reports. We draw on interviews with 41 participants
in High River (n = 35) and Calgary (n = 6). In addition to residents, the High River
case study captures interviews with decision-makers and those in advisory roles who
are representatives of governments, organisations, scientific institutions, media and
the private sector involved in flood management at the municipal, regional, provincial and federal levels. Interviews for the Calgary case study (C-CS) were conducted
in spring 2014 and the High River interviews (HR-CS) were conducted in spring
2015; both within the 24-month critical window of opportunity after the flood event.

Findings and discussion

In the face of the unprecedented disaster, both Calgary and High River have recovered relatively well with assistance from provincial and federal governments. Calgary
and High River have worked tirelessly to reach their current stages of recovery; we
provide only brief descriptions of their efforts. Both municipalities' recovery efforts
aimed for resiliency and civic involvement. The following discussion compares and
contrasts the extent to which the two municipalities achieved these aims.

Disaster recovery overview

The damage and the scope of recovery varied in the two cases: Calgary suffered
only partial inundation while High River experienced town-wide inundation.
The difference in the return of residents to their homes and businesses created
further significant differences in recovery of the two municipalities. On 20 June
2013, Mandatory Evacuation Orders were announced for some neighbourhoods
in Calgary and for all residents of High River. Though controversial at the time
due to safety concerns, Calgarians were given permission by Mayor Naheed Nen-
shi to return to their homes and businesses within three days of the flood event.

In contrast, High River residents were not allowed to return for two weeks or
more, causing them to be on the "verge of some very major civil disobedience"
(Richards and Howell, 2013, para. 11). The time difference in returning to one's
home or business is significant for three reasons: 1) flood damage requires quick

clean-up to minimise damage and to reduce mould growth; 2) businesses provide jobs which in turn provide financial resources critical for rebuilding and recovery; and 3) community participation and taking action reduce anxiety and trauma and empower people by increasing sense of control (Johnston *et al.*, 2012).

The City of Calgary convened a Task Force to oversee municipal recovery efforts. Calgary's municipal planners identified "the rebuilding timeframe is a critical window of opportunity where citizens feel a *sense of belonging* and are keen to make the future better through *collective action*" (Arthurs, 2015, para. 22; italics added). In addition, the City recognised that "community recovery must not only address the vitality of the built, economic, natural, and social environments but also reduce the risk of future disaster events in order to build a more *disaster-resilient community*" (CoC, 2016, para. 8; italics added). To achieve this goal, the City would need to explore ways to "develop and sustain *long-term resiliency*" (ibid.; italics added) and to use "a *holistic, community-based approach*" (ibid., para. 9; italics added).

In the Town of High River, a Renew Committee was developed to create a long-term recovery team to get people back into damaged homes. The Town's vision was "High River is a people first community on the Highwood River where we live, work and play" (ToHR, 2015b, p. 5), where the community is connected, including through "*active community engagement*" (ibid., p. 9).

Public engagement for supporting status quo ante or long-term sustainability?

Numerous stakeholder engagement sessions on flood mitigation were held between 2013 and 2015. Community leaders and decision-makers, such as mayors and councillors, industry and NGOs, were involved extensively in regional recovery coordination and community flood mitigation planning. However, feedback from the general public was mostly limited to three formats: 1) open house-type design to inform or decide–announce–defend with props such as posters and maps where officials or experts answered specific questions from residents; 2) feedback on reports via websites; and 3) feedback via social media.

Largely, community engagement about flooding did not create *a sense of belonging* or *collective action* for our research participants. Community engagement events either focused on individual, short-term flood preparedness and supporting individual resilience through information sessions with government experts (e.g. provincial flood information sessions, 23–24 September 2013) or served as platforms for expert presentations (e.g. Provincial Flood Symposium, 4 October 2013) with controlled question-and-answer periods. Although presented as an event that would bring "together experts, community representatives and Albertans to discuss ideas and best practices for flood mitigation" (Government of Alberta, 2013b), "the symposium was not a symposium at all . . . little dialogue was encouraged, and when it was permitted it was constrained in a way that did not countenance penetrating questions of differing points of view" (Sandford and Freek, 2014, p. 50). The description of this event matches our participants' experiences of other public engagements regarding the flood.

Calgary

Numerous community meetings and open houses were held in Calgary and were attended by more than 6,000 residents in the first year (CoC, 2014b). However, the public venues for engagement were sometimes bleak, utilitarian and focused almost exclusively on individual flood relief and prevention. In one case, during Alberta's Watershed Management Symposium: Flood and Drought Mitigation (29 April 2014), residents were literally left to sit in the dark and their only engagement opportunity was to tweet their questions to a revolving roster of scientific experts or speak one-on-one with representatives in the back of the room. Controversial flood mitigation options for the Calgary region—Springbank off-stream storage, McLean Creek dry dam and diversion tunnel—were presented in an open-house format because of the range of conflicting views and opinions of residents and property owners from various areas around the region.

While supporting residents to return to daily functioning, the focus on individual resilience and expert solutions diminished discussions of community action from flood recovery. Interview participants either did not feel they belonged at the engagement events or were disappointed by the process. For example:

> They weren't answering people's questions. They were pussyfooting around the topic. And . . . the questions they [the public] were asking were intelligent questions. They wanted intelligent responses. And they didn't get that from the City people, or the Provincial people, or some expert on flooding, and stuff like that. They just didn't get it. And it turned into . . . very upsetting for some people I think.
>
> *(C-CS, 2014)*

One Calgary resident felt the only way to provide input was to write letters: "to the Premier of Alberta, the Mayor of Calgary and my Allstate Insurance agent regarding the failure of provincial, city, insurance and banking authorities to keep development out of the flood plain" (HR-CS, 2015).

Frustrations were expressed about attending public engagement events if other attendees had lifestyles or made choices that conflicted with sustainability:

> I think there's all these meetings, flood prep and flood mitigation, flood disaster and . . . I don't want to hear hope to people crying and bitching, like from their glass houses. I don't mean to sound insensitive . . . And they want answers. And people want to talk about it . . . But other people don't want to talk about it. And I do want to talk about it, like I said I hope they fix the sewers this time. And to live in a flood plain and not expect it to ever flood, are you freaking kidding me?
>
> *(C-CS, 2014)*

This participant's observation exemplifies the problems when public engagement turns into an unproductive, time-consuming and costly process that overemphasises special interests and lacks representation. For example, the Calgary River Communities Action Group (CRCAG) was formed by residents in neighbourhoods located in flood hazard areas, and one of their mandates (No. 5) is to "advocate against policy, legislative and regulatory changes that are prejudicial to and negatively impact property owners in flood impacted communities" (CRCAG, 2015, para. 13). In other words, they are resisting the provincial government's plans to relocate homes via the provincial floodway buyout programme and are fighting for extensive flood mitigation infrastructure to protect their homes, which would be funded by taxpayers. Many participants do not feel this arrangement of risk apportionment is fair, as expressed here:

> You get these claims coming in for reimbursement by the province and the insurance companies. My house insurance goes up by a factor of two roughly. Why am I paying insurance for someone to live in a nice location? . . . Most of the people in Calgary should complain about the flood mitigation stuff going on because they're paying for a few people to be looked after . . . But why should everybody pay for that?
>
> *(HR-CS, 2015)*

Those disappointed by the municipal and provincial engagement processes proposed alternate options for engagement to be more inclusive, integrating, participatory and *forward looking*:

> And I just don't know what sandbags are gonna do, you know? Is it just that we build this huge fortress of sandbags, does that mean all our neighbours are gonna get hammered by it? I don't know. So to get to become more active in the community and join the Community Association and get to understanding of what, what we can do as the community, as a city, to prevent this from happening again.
>
> *(C-CS, 2014)*

It is important to note that these interviews took place nine months after the flood event. While participants were still surfacing and dealing with the trauma they had experienced, they were also nonetheless looking for opportunities to use their experience for greater engagement and social change.

High River

The State of Local Emergency was not lifted until about three months after the flood and consultation was limited during the early stages. Due to flood damage, the Town's municipal infrastructure and facilities were not functioning, creating safety concerns and limiting the availability of public spaces to gather. Since residents were scattered throughout the province, and home and business renovations were delayed, many people did not have the emotional and physical capacity to attend public consultations in the immediate aftermath. The Town's administration, mayor and some councillors began using traditional and social media to communicate with residents near and far. Their Facebook page is still used extensively by residents to provide negative and positive feedback. While a communication platform in its own right, it cannot replace the face-to-face community dialogue with targeted discussion on re-visioning the future and collective action.

Once the Town was functional again, town hall meetings were set up to engage the public in flood recovery process. These meetings became a venue for furious residents to vent about not being allowed back to their properties for weeks, about the unlawful seizure of secured firearms by police (Government of Canada, 2015) and about forceful entry of homes by knocking down doors and leaving properties vulnerable to theft and further damage (Weismiller, 2013). Public engagement in High River "was basically anger management sessions" (HR-CS, 2015).

Given the emotional intensity of the town hall meetings, public engagements thereafter were designed as open houses, which, as discussed earlier, do not provide opportunities for meaningful participation or dialogue and for visioning collective action and alternate futures. A participant assessed the purpose of open houses in the following way:

> I think the open houses were more to educate them [the public] and make them feel a part of the process, so they understood what was happening. Different than other open houses, where you're looking for feedback and looking at options, etc. . . . I think the plans that were put in place were so obvious that they needed to be done . . . It wasn't a debatable issue. It wasn't going to cause harm to our local residents. It was going to protect them.
>
> *(HR-CS, 2015)*

Residents were asked for feedback regarding some projects, such as new design ideas for downtown, but not projects requiring technical expertise such as dike systems. An interviewee explained:

> Our goal wasn't so much to get feedback from people about where dikes should go and where they should be built, as it was to say, 'Look. Here's what we're building and we want to give you an opportunity to speak to people.' Because it's kind of like you going and giving advice to your mechanic, in some ways, right? Would you give advice to your mechanic?
>
> *(HR-CS, 2015)*

While some decisions require expert knowledge, lay knowledge can also provide valuable input. Moreover, learning from one another can lead to "creating or agreeing on a new understanding of the world that incorporates new perspectives" (Hoverman *et al.*, 2011, p. 29) and can facilitate collective action towards sustainability goals.

From the perspective of those organising public engagements, providing opportunities for citizens' voices was not necessarily seen as constructive or representative:

> Open houses, you only get the people that have a gripe about what you were doing. Regardless, of what you were doing . . . We had those type of people

and they're the ones that became very vocal in the community, it was a small minority over the majority.

(HR-CS, 2015)

Open-house formats are not welcoming (especially for those who have been disappointed in the past) and do not allow for two-way, in-depth dialogue that is more common in workshops and working groups. The challenge of engaging citizens and prompting participation from a range of stakeholders and a majority rather than a minority, noted in the quote above, is a common theme raised by governments in Canada and elsewhere regarding issues beyond disasters. Some jurisdictions in Canada and Europe have found approaches that are successful (see 'Literature review' on p. 39). Other types of public engagement formats were designed to "engage and provide opportunities for residents to get together" (ToHR, 2015a, p. 13) at events such as teas, luncheons, flood anniversary gatherings, parades, shows, carnivals, etc. These types of social events create spaces for dialogue between residents and play an important function; however, they should not be used as substitutes for public engagement events that facilitate dialogue between and among residents and decision-makers in ways that enhance social learning on complex flood risk management issues and joint decision-making.

In summary, the engagement opportunities in our case studies either focused on individual, short-term flood preparedness and individual resilience or were platforms for expert presentations. Despite municipal and provincial government objectives to support resiliency and involve residents, these formats prevented public dialogue on broader issues that case study participants raised: lack of in-depth, two-way dialogue; lack of fairness and apportionment of responsibility for inappropriate land-use and development; lack of provincial regulations or enforcement of development in flood-prone areas in the past; and over-representation of particular interests. While many of the structural and non-structural measures implemented since the 2013 flood were designed to minimise or prevent future flood damage, our data provides evidence of residents expressing an interest in more extensive engagement, exploration of issues and empowerment.

Our findings indicate that in the early stages of recovery, authorities were not well prepared to deal with the highly charged emotional atmosphere of community events. To do so would have required formats other than large open houses and a variety of engagement tools. In the face of inevitable loss and change, planners and resource managers often struggle to find effective approaches to engaging their communities and sometimes shy away from collective community visioning due to concerns about stakeholders' responses (Moser, 2012). In post-disaster contexts, the stakes and tensions are high, the decision timeframes are short, and public and political pressure call for quick decisions, solutions and investments in returning back to 'normal'. While these conditions create formidable challenges, more meaningful public engagement around flood management issues in Alberta—including during recovery—is both necessary and possible, as exemplified in other jurisdictions in Canada and around the world.

Conclusion

Since many disasters originate from social processes of unsustainable practices, risks need to be socially negotiated. Meaningful public engagements enable people to create collective understanding of risks and vulnerabilities and to shift from individual interests to more civic-minded approaches to addressing complex socio-environmental issues. Based on our findings, the format of public engagements in Calgary and High River following the 2013 floods was mostly limited to one-way information dissemination. Such formats minimised participatory processes and opportunities for dialogue about causal factors such as building in flood-hazard areas and also did not create a sense of belonging and collective action. While community authorities may have been aware of the psychological phases of disaster response, in practice our findings suggest that they were not incorporated into the timing or approaches to community consultation. In the future, a variety of approaches that create safe spaces for dialogue and social learning and that are supportive and reflective of the emotional phases of recovery would be advantageous as part of the pre-disaster recovery planning process.

The 2013 changes in the Alberta *Municipal Government Act* (*Bill 27*; see Alberta Municipal Affairs, 2015) enables regulation-making powers to limit floodway development. Currently, the regulations and bylaws have yet to be promulgated, meaning that there has, in effect, been no regulatory change since 2013. Further, political motivations respond to powerful interests. In the absence of substantive transformative change at the societal level, vested interests may seek to undo the regulatory and legislative changes enacted in the aftermath of the flood. The public engagement strategies described in this chapter did not achieve the full potential for *forward-looking* recovery following the 2013 Alberta flood. Therefore, these strategies inhibited a *cultural shift* to "disaster risk reduction as a *way of life*" (Public Safety Canada, 2016, p. 3; italics added) and limited opportunities for building "resilient capacity", which necessitates a process that would *empower* all members of society to *share responsibility* for reducing vulnerability and keeping hazards from becoming disasters (Public Safety Canada, 2011).

Two of the limitations of this study are the small number of participants (n = 41) and that participants as well as we, the researchers, did not participate in all of the public engagement events after the flood. As such, our evaluation of public engagement is limited to those events we, or our participants, attended and those described in reports and the media. Despite these limitations, our research provides sufficient evidence to highlight opportunities for improvement in public engagement during post-disaster recovery in Alberta, where public involvement in natural resource management issues has tended to lag behind other jurisdictions.

This research contributes to an improved understanding of place-based disaster recovery and input into the development of recovery policies and practices that achieve transformative change and community resiliency. Our recommendations include the following: 1) further examine case studies in other jurisdictions with a general culture of more collaborative in-depth community engagement and the lessons learned for flood risk management; 2) when designing community engagement formats, create conditions that lead to meaningful engagement and create a multifaceted strategy that considers various psychological phases; and 3) seek opportunities during the recovery stage to build resilient capacity and collective action for disaster risk reduction. In moving to more meaningful engagement, Albertans in general would gain a greater understanding of individual and collective risk apportionment, thereby facilitating a cultural shift and building the social capacity needed for actions that lead to long-term reduction in overall flood risk.

Notes

1 A review of 15 years of research on learning for sustainability by Sinclair *et al.* (2008) concluded that characteristics of meaningful engagement include: early and inclusive involvement that creates opportunities for identifying and resolving conflicts over diverse norms, values and aspirations; deliberative involvement that stimulates dialogue and development of collaborative relations; and empowerment that fuels a sense of agency and socio-political action.
2 Social learning is characterised by an iterative process, reflective practice, utilisation of diversity, shared understanding and experimentation (Rodela *et al.*, 2012).
3 GoA's stakeholder engagement summaries for various river basin studies can be found at www.alberta.ca/flood-mitigation-studics.cfm.

References

Alberta Municipal Affairs (2015) *Overview of Bill 27, Floodway Development Regulation Consultation*. Edmonton: Government of Alberta. Available at: www.municipalaffairs.alberta.ca/1934 (accessed 16 September 2015).

Alexander, D.E. (2013) 'Resilience and Disaster Risk Reduction: An Etymological Journey', *Natural Hazards and Earth System Science*, 13(11), pp. 2707–2716.

Arthurs, C. (2015) 'Recover, Repair, Prepare: Calgary after the 2013 Flood' World Conference on Disaster Management (WCDM) Connect blog, 20 February. Available at: www.wcdm.org/blog/recover-repair-prepare-calgary-after-the-2013-flood.html (accessed 28 February 2016).

Ashley, R.M., Blanskby, J., Newman, R., Gersonius, B., Poole, A., Lindley, G., Smith, S., Ogden, S. and Nowell, R. (2012) 'Learning and Action Alliances to Build Capacity for Flood Resilience', *Journal of Flood Risk Management*, 5(1), pp. 14–22.

Auditor General (2015) *Report of the Auditor General of Alberta*. Edmonton: Government of Alberta, March. Available at: www.oag.ab.ca/webfiles/reports/OAG%20 March%20 2015%20Report.pdf (accessed 22 February 2016).

Berke, P.R., Kartez, J. and Wenger, D. (1993) 'Recovery after Disaster: Achieving Sustainable Development, Mitigation and Equity', *Disasters*, 17(2), pp. 93–109.

Bogdan, E.A. (n.d.) 'Flooding Discourse: Perceptions and Practices of Flood Management in High River, Alberta', unpublished PhD thesis, University of Alberta.

CoC (2014a) 'Civic Census Results'. City of Calgary. Available at: www.calgary.ca/CA/city-clerks/Documents/Election-and-information-services/Census2014/Final%20 2014% 20Census%20Results%20book.pdf (accessed 5 September 2016).

CoC (2014b) 'Calgary Recovers Building for Resiliency'. City of Calgary. Available at: www.calgary.ca/General/flood-recovery/Pages/Calgary-flood-2013-infographic-recap.aspx (accessed 5 September 2016).

CoC (2016) 'Flood Recovery'. City of Calgary. Available at: www.calgary.ca/general/flood-recovery/Pages/FloodRecoveryHome.aspx (accessed 5 September 2016).

Cox, R.S. and Perry, K.M.E. (2011) 'Like a Fish out of Water: Reconsidering Disaster Recovery and the Role of Place and Social Capital in Community Disaster Resilience', *American Journal of Community Psychology*, 48(3–4), pp. 395–411.

Craig, D. (1990) 'Social Impact Assessment: Politically Oriented Approaches and Applications', *Environmental Impact Assessment Review*, 10, pp. 37–54.

CRCAG (2015) *AGM Summary and Board Comments*. Calgary River Communities Action Group, 9 December. Available at: http://protectcalgary.com/agm-summary-and-board-comments (accessed 20 February 2016).

DeWolfe, D.J. (2000) *Training Manual for Mental Health and Human Service Workers in Major Disasters*, 2nd edn. HHS Publication No. ADM 90–538. Rockville, MD: US Department of Health and Human Services, Substance Abuse and Mental Health Services Administration, Center for Mental Health Services.

Diduck, A. and Mitchell, B. (2003) 'Learning, Public Involvement and Environmental Assessment: A Canadian Case Study', *Journal of Environmental Assessment Policy and Management*, 5(3), pp. 339–364.

Freudenburg, W.R., Gramling, R.B., Laska, S. and Erikson, K. (2009) *Catastrophe in the Making: The Engineering of Katrina and the Disasters of Tomorrow*. Washington, DC: Island Press.

Gordon, R. (2008) *A "Social Biopsy" of Social Process and Personal Responses in Recovery from Natural Disaster*. GNS Science Report 2008/09, February. Lower Hutt: Institute of Geological and Nuclear Sciences Ltd.

Government of Alberta (2013a) *Southern Alberta 2013 Floods: The Provincial Recovery Framework*. Available at: http://alberta.ca/albertacode/images/Flood-Recovery-Framework.pdf (accessed 8 July 2016).

Government of Alberta (2013b) *The Alberta Flood Mitigation Symposium*. Available at: www.alberta.ca/flood-symposium.cfm (accessed 12 February 2016).

Government of Canada (2015) *Chair-Initiated Complaint and Public Interest Investigation into the RCMP's Response to the 2013 Flood in High River, Alberta*. Available at: www.crcc-ccetp.gc.ca/en/chair-initiated-complaint-and-public-interest-investigation-rcmps-response-2013-flood-high-river (accessed 12 February 2016).

Haque, C.E., Kolba, M., Morton, P. and Quinn, N.P. (2002) 'Public Involvement in the Red River Basin Management Decisions and Preparedness for the Next Flood', *Global Environmental Change Part B: Environmental Hazards*, 4(4), pp. 87–104.

Hayward, G., Diduck, A. and Mitchell, B. (2007) 'Social Learning Outcomes in the Red River Floodway Environmental Assessment', *Environmental Practice*, 9(4), pp. 239–250.

Henstra, D. and McBean, G. (2005) 'Canadian Disaster Management Policy: Moving toward a Paradigm Shift?', *Canadian Public Policy*, 31(3), pp. 303–318.

Hoverman, S., Ross, H., Chan, T. and Powell, B. (2011) 'Social Learning through Participatory Integrated Catchment Risk Assessment in the Solomon Islands', *Ecology and Society*, 16(2), pp. 17–39. Available at: www.ecologyandsociety.org/vol16/iss2/art17 (accessed 28 February 2016).

IBI Group (2015) *Provincial Flood Damage Assessment Study*. Available at: http://aep.alberta.ca/water/programs-and-services/flood-mitigation/documents/pfdas-alberta-main.pdf (accessed 14 February 2016).

Johnston, D., Becker, J. and Paton, D. (2012) 'Multi-Agency Community Engagement During Disaster Recovery: Lessons from Two New Zealand Earthquake Events', *Disaster Prevention and Management: An International Journal*, 21(2), pp. 252–268.

Kim, K. and Olshansky, R. (2014) 'The Theory and Practice of Building Back Better', *Journal of the American Planning Association*, 80(4), pp. 289–292. Available at www.tandfonline.com/doi/abs/10.1080/01944363.2014.988597 (accessed 28 February 2016).

Kreimer, A. (1978) 'Post-Disaster Reconstruction Planning: The Cases of Nicaragua and Guatemala', *Mass Emergencies*, 3(1), pp. 23–40.

Lertzman, R. (2014) *Psychosocial Contributions to Climate Sciences Communications Research and Practice*. Available at: www.ucl.ac.uk/public-policy/policy_commissions/Commu nication-climate-science/Communication-climate-science-report/psychosocial_final. pdf (accessed 1 April 2016).

Macy, J. and Johnstone, C. (2012) *Active Hope*. Novato, CA: New World Library.

McCarthy, D.D., Crandall, D.D., Whitelaw, G.S., General, Z. and Tsuji, L.J. (2011) 'A Critical Systems Approach to Social Learning: Building Adaptive Capacity in Social, Ecological, Epistemological (SEE) Systems', *Ecology and Society*, 16(3), pp. 18–34.

Mileti, D. (1999) *Disasters by Design: A Reassessment of Natural Hazards in the United States*. Washington, DC: Joseph Henry Press.

Mnguni, P.P. (2010) 'Anxiety and Defense in Sustainability', *Psychoanalysis, Culture and Society*, 15, pp. 117–135.

Moser, S. (2012) 'Getting Real about It: Navigating the Psychological and Social Demands of a World in Distress'. In D.R. Gallagher, R.N.L. Andrews and N.L. Christensen (eds) *Sage Handbook on Environmental Leadership*. Thousand Oaks, CA: Sage, pp. 432–440.

Nicholsen, S.W. (2003) *The Love of Nature and the End of the World*. Cambridge, MA: MIT Press.

Olshansky, R.B. and Chang, S. (2009) 'Planning for Disaster Recovery: Emerging Research Needs and Challenges', *Program Planning*, 72(4), pp. 200–209.

Pearce, L. (2003) 'Disaster Management and Community Planning, and Public Partici- pation: How to Achieve Sustainable Hazard Mitigation', *Natural Hazards*, 28(2–3), pp. 211–228.

Pelling, M. (2003) *The Vulnerability of Cities: Natural Disasters and Social Resilience*. London: Earthscan.

Perry, R.W. (2007) 'What Is a Disaster?' In H. Rodríguez, E.L. Quarantelli and R.R. Dynes (eds) *Handbook of Disaster Research*. New York: Springer, pp. 1–15.

Pomeroy, J.W., Stewart, R.E. and Whitfield, P.H. (2016) 'The 2013 Flood Event in the South Saskatchewan and Elk River Basins: Causes, Assessment and Damages', *Canadian Water Resources Journal*, 41(1 2), pp. 105 117.

Public Safety Canada (2011) *An Emergency Management Framework for Canada*, 2nd edn. Available at: www.publicsafety.gc.ca/cnt/rsrcs/pblctns/mrgnc-mngmnt-frmwrk/index- eng.aspx#a06 (accessed 25 February 2016).

Public Safety Canada (2016) *Canada National Disaster Mitigation Strategy*. Available at: www. publicsafety.gc.ca/cnt/rsrcs/pblctns/mtgtn-strtgy/index-eng.aspx (accessed 25 February 2016).

Randall, R. (2009) 'Loss and Climate Change: The Cost of Parallel Narratives', *Ecopsychol- ogy*, 1, pp. 118–129.

Richards, G. and Howell, T. (2013) 'High River Residents Demand to Return Home', *Calgary Herald*, 26 June. Available at: www.calgaryherald.com/High+River+residents+demand+ return+home/8584527/story.html (accessed 25 September 2015).

Rodela, R., Cundill, G. and Wals, A.E. (2012) 'An Analysis of the Methodological Under- pinnings of Social Learning Research in Natural Resource Management', *Ecological Eco- nomics*, 77, pp. 16–26.

Rubin, C. and Barbee, D. (1985) 'Disaster Recovery and Hazard Mitigation: Bridging the Intergovernmental Gap', *Public Administration Review*, 45, pp. 57–63. Available at: www. jstor.org/stable/3134998 (accessed 28 February 2016).

Sandford, R. and Freek, K. (2014) *Flood Forecast: Climate Risk and Resiliency in Canada*. Toronto: Rocky Mountain Books.

Shrubsole, D. (2013) 'A History of Flood Management Strategies in Canada Revisited'. In E.C.H. Keskitalo (ed.) *Climate Change and Flood Risk Management: Adaptation and Extreme Events at the Local Level*. Cheltenham: Edward Elgar, pp. 95–120.

Sinclair, A.J. and Diduck, A.P. (2001) 'Public Involvement in EA in Canada: A Transformative Learning Perspective', *Environmental Impact Assessment Review*, 21(2), pp. 113–136.

Sinclair, A.J., Diduck, A. and Fitzpatrick, P. (2008) 'Conceptualizing Learning for Sustainability through Environmental Assessment: Critical Reflections on 15 Years of Research', *Environmental Impact Assessment Review*, 28(7), pp. 415–428.

Smith, G.P. and Wenger, D. (2007) 'Sustainable Disaster Recovery: Operationalizing an Existing Agenda'. In H. Rodríguez, E.L. Quarantelli and R.R. Dynes (eds) *Handbook of Disaster Research*. New York: Springer, pp. 234–257.

Spee, K. (2008) *Community Recovery after the 2005 Matata Disaster: Long-Term Psychological and Social Impacts*. GNS Science Report 2008/12. Lower Hutt: Institute of Geological and Nuclear Sciences Ltd.

Statistics Canada (2014) *NHS Profile, High River, CA, Alberta, 2011*. Available at: www 12.statcan.gc.ca/nhs-enm/2011/dp-pd/prof/details/page.cfm?Lang=EandGeo1=CMA andCode1=821andData=CountandSearchText=High%20RiverandSearchType= BeginsandSearchPR=01andA1=AllandB1=AllandTABID=1 (accessed 11 September 2015).

Tierney, K.J. (2012) 'Disaster Governance: Social, Political, and Economic Dimensions', *Annual Review of Environment and Resources*, 37, pp. 341–363.

ToHR (2015a) *Report to Citizens on Renewal Activities*. Town of High River. Available at: www.highriver.ca/images/Communications/2015/2015_report-to-citizens_02-19-15_ web.pdf (accessed 6 September 2015).

ToHR (2015b) *Town of High River Strategic Plan*. Town of High River. Available at: http:// highriver.ca/images/Leg_Services/corporate-strategic-plan_2015_-_2017.pdf (accessed 6 September 2015).

Vallance, S. (2011) 'Early Disaster Recovery: A Guide for Communities', *Australasian Journal of Disaster and Trauma Studies*, 2, pp. 19–25.

van Herk, S., Zevenbergen, C., Ashley, R. and Rijke, J. (2011) 'Learning and Action Alliances for the Integration of Flood Risk Management into Urban Planning: A New Framework from Empirical Evidence from the Netherlands', *Environmental Science and Policy*, 14(5), pp. 543–554.

Ward, J., Becker, J. and Johnston, D. (2008) *Community Participation in Recovery Planning: A Case Study from the 1998 Ohura Flood*. GNS Science Report 2008/22. Lower Hutt: Institute of Geological and Nuclear Sciences Ltd.

Webb, G.R. (2007) 'The Popular Culture of Disaster: Exploring a New Dimension of Disaster Research'. In H. Rodríguez, E.L. Quarantelli and R.R. Dynes (eds) *Handbook of Disaster Research*. New York: Springer, pp. 430–440.

Weintrobe, S. (2013) 'The Difficult Problem of Anxiety in Thinking about Climate Change'. In S. Weintrobe (ed.) *Engaging with Climate Change: Psychoanalytic and Interdisciplinary Perspectives*. London: Routledge, pp. 33–47.

Weismiller, B. (2013) 'High River Residents Vent Anger at Meeting', *Calgary Herald*, 31 July. Available at: www.calgaryherald.com/sports/High+River+residents+vent+anger+ meeting/8733951/story.html (accessed 15 December 2015).

5

HOW THE CHINESE GOVERNMENT RESPONDED TO THE WENCHUAN EARTHQUAKE

Yung-Fang Chen

Introduction

A magnitude 7.9 earthquake struck eastern Sichuan, China, on 12 May 2008 at 06:28. The impact areas included Sichuan, Shaanxi and Gansu provinces, within which 51 counties were affected. The main impact areas that were hit are indicated in Figure 5.1. The earthquake resulted in 69,266 deaths, 374,643 injured and 17,923 missing (UNCRD 2009). More than 15 million people were affected and had to be evacuated from their original homes (Gu *et al.* 2011). The total financial loss from the earthquake was RMB845.1 billion (US$128.4 billion) (UNCRD 2009).

Just prior to this event, the Chinese government had passed the *Civil Contingency Act 2007* to provide a national framework for emergency management. Consequently, on the one hand, the government wanted to demonstrate its capacity to respond to such a large catastrophe to fulfil nationalist self-esteem and self-confidence after the recent socioeconomic development (Xu and Lu 2008). However, on the other hand, the Chinese government soon realised that the challenges for reconstruction would not merely focus on housing and livelihood reconstruction. Despite the normal emergency management framework in China being heavily based on a centralised system, the response and recovery efforts were provided by local authorities and provincial governments (Hu *et al.* 2010). It is also argued that the responsibility of responding to and recovering from disasters have long been responsibilities of the general public (Otani 2014). At the time, there was not yet sufficient regional and/or provincial support for disaster recovery (Ge *et al.* 2010). This lack of clarity in the command and control system prevented response and recovery systems from working effectively after the Wenchuan earthquake.

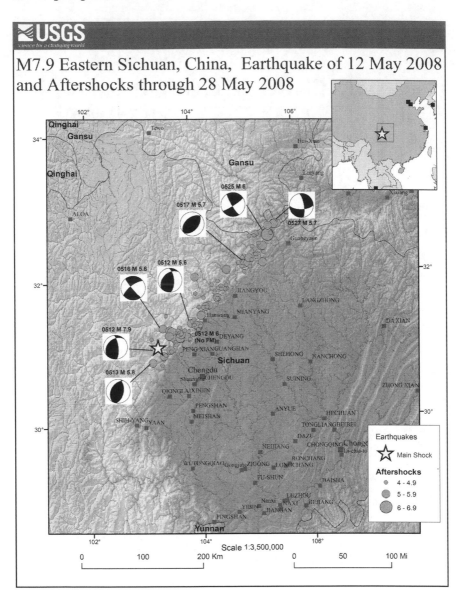

FIGURE 5.1 Map showing the location of the 2008 Sichuan earthquake and all the aftershocks following it up to 28 May 2008.

Source: USGS 2008.

The national reconstruction strategies post-Wenchuan earthquake

To respond to such a large disaster, the Chinese government initiated several regulations and plans for recovery and reconstruction for the sites (State Council

2008a). Based on the newly enacted *Civil Contingency Act* (2007) and *Law of Protecting against and the Mitigation of Earthquake Disasters* (1997), the State Council announced the *One on One Assistance Programme Post Wenchuan Earthquake* on 4 June (State Council 2008b). The finalised framework for reconstruction was summarised on 11 August 2008 in *The State Overall Planning for Post Wenchuan Earthquake Restoration and Reconstruction* 2008 (PGPWERR 2008; see also State Council 2008c) to provide strategies and milestones for the reconstruction work.

The plan not only stressed the priority of reconstructing residential houses and public facilities, it also encouraged the local authorities to use the reconstruction as a long-term sustainable development opportunity, and to maximise the chance to facilitate economic growth, through rural and ecosystems development (State Council 2008c). The concept of 'build back better' was included in the sustainable development plans, since it not only included recovery, but also the mitigation of and preparation for disasters (Otani 2014; Hu *et al.* 2010: 111). The development plans were tailored to the government's existing 'Development of the West' policy and 'scientific development' approaches (Wei *et al.* 2012).

What is noted is that the plans utilised a 'one-on-one assistance approach' to facilitate effective recovery and reconstruction (State Council 2008b). The 18 counties (and their cities) seriously affected by the earthquake in Sichuan, Gansu and Shaanxi provinces (assistance receivers) were paired with 20 comparably developed provinces (and their cities) elsewhere in China (assistance providers). Following the centralised design principle, these twinned assistance providers were asked to utilise 1 per cent of their annual gross domestic product (GDP) as funding to provide support for development for a period of three years (EERI 2008; Hu *et al.* 2010; Wei *et al.* 2012). The State Council promised that RMB 1 trillion (US$157 billion) would be allocated for reconstruction of the affected areas. By May 2011, 95 per cent of the projects had been completed and RMB 885 billion (US$138 billion) had been spent (Feng *et al.* 2016: 12).

Scholars have attempted to prove that the 'top-down' government-led co-ordination of post-Wenchuan earthquake reconstruction has been effective. Kamata (2011) argued that the Chinese government raised the speed of recovery and reconstruction to demonstrate its ruling power after the earthquake struck; hence, the reconstruction plan was implemented much earlier than with other disasters (Kamata 2011). The top-down post-disaster planning and management mechanism enabled the government's intervention in financial aid, provision of relief goods and setting of priorities to be more effective (Ge *et al.* 2010). Xu and Lu (2008) summarised the programme as being suitable for a developing country, since the programme facilitated the best possible mobilisation of national resources from developed areas to affected areas; in addition, by using the resources from developed provinces, the programme reduced the reliance on international aid, and reduced the pressure on the central government (Xu and Lu 2008). Criticisms include accusations that the generalised state policy showed a lack of flexibility and that it was unconcerned with the specifics of local needs and characteristics (Ge *et al.* 2010). Neither did the programme pay attention to social recovery, such as resettlement, production resumption, and livelihood and social order (Xu and Lu 2008). A further complicating issue was that many local government officers and their teams were missing or dead after the earthquake, which paralysed the effectiveness of public services delivery in the recovery phase (Li and Lin 2013).

It was the first time that the Chinese government had recognised the usefulness of, and had involved, NGOs and local resources to support a reconstruction programme (Hirai 2009: I-16; Hu *et al.* 2010: 111). The involvement of NGOs meant that they could respond to the needs of the affected populations and could act as communication channels between the government and the public (Lu and Xu 2014: 268); some NGOs also used the opportunity to be involved in the planning procedure (World Economic Forum 2010). The NGOs and volunteers were mostly involved in roles of coordination and implementation, while the government acted as supporter and policy maker (Zhang and Yu 2009). Although there were some criticisms regarding a lack of appropriate qualifications, capacity and competences of the social workers from the NGOs during the reconstruction phase (Huang *et al.* 2014), the role of NGOs in Chinese society was recognised (Li and Lin 2013; Otani 2014). The Chinese government even published an official white paper to suggest that the government should enhance the 'social mobilisation mechanisms', so that NGOs and volunteers could play a greater and fuller role in the field of 'disaster prevention, emergency rescue, relief and donation work, medical, hygiene and quarantine requirements, post disaster reconstruction, psychological support and other tasks' (State Council, quoted in Roney 2011: 85). The change in the role of NGOs in emergency management could be seen in the revised Article 7.5.1 in the *Civil Contingency Act* (2016). The article suggests that NGOs and volunteers are encouraged to be trained to be able to participate in disaster response (State Council 2016).

The impact of the Wenchuan earthquake on Beichuan County

Among the 20 seriously affected areas, the reconstruction in Beichuan County has attracted most attention, not only because it was the worst-hit county, but also because of the special geographical characteristics and its demographic composition. Old Beichuan was located in a hilly area 600 metres above sea level, with the Anchuang River and the Beichuan–Yingxiu fault line running thorough the county. The high mountains and steep valleys result in high risks of landslides, mudslides and other types of geological disaster (Zhu *et al.* 2012). Old Beichuan County was comprised of 3 towns, 17 districts and 278 villages. The population before the earthquake was 1.61 million. Minority ethnic groups such as the Qiang, Zang and Hui tribes accounted for 50.5 per cent of the total population (CAUPD 2008). Since the Qiang tribe has resided in old Beichuan for a long time, accounting for 95.8 per cent of its minority ethnic groups, the State Council made Beichuan the Qiang Autonomous County in 2003.

FIGURE 5.2 View of Qushan Town post-Wenchuan earthquake.

Old Beichuan County was swept aside by a massive landslide, and 28.98 per cent of the overall population was impacted by the earthquake. This included 15,645 deaths, 26,916 injuries and 4,402 missing (CAUPD 2008). Almost 80 per cent of the buildings in the county were seriously damaged, together with major damage to roads, bridges and other infrastructure (Liu *et al.* 2011: 927), as well as schools and public services. Taking as an example Qushan Town, where the county council was located, it was buried by a large-scale collapse and landslide after the earthquake. Figure 5.2 shows a view of the town post-Wenchuan earthquake. The bottom three floors of the buildings in the photograph are all underground.

The strategies of reconstruction in Beichuan County

Prime Minister Wen gave instructions to create a new Beichuan and paid personal attention to reconstruction in this county. Shandong Province was selected to support the reconstruction in the one-on-one assistance programme. Following the guidelines from the State Council, the Department of Development and Reform Commission in the Sichuan provincial government played the role of coordinating the aid providing province (officers and technical workers from Shandong) with the aid receiving county (officers and public from Beichuan) (Xu and Lu 2008). The China Academy of Urban Planning and Design (CAUPD) was designated the chief institute of design and planning for the (re)construction of the affected areas in Beichuan (Zhu *et al.* 2012).

Based on the PGPWERR (2008), the one-on-one assistance programme started with an initial impact assessment, so that recommendations could be provided for recovery and reconstruction plans. Although on-site reconstruction was more favourable, as many as 35,438 of the affected population required a relocation plan (see Table 5.1); most of them were relocated to another location/town. The limited amount of land available meant that it was more difficult to find suitable land to cope with the high numbers of affected people. After consultation with experts and affected people, old Anchuang (currently renamed as Yongchang, but also known as 'New Beichuan') was assigned as a new construction site to take most of the affected people. Anchuang (Yongchang/New Beichuan) is situated about 23 kilometres south-east of old Beichuan.

The 'government-led public participation' approach encouraged the affected population not just to receive relevant information passively, but to be proactively involved in the discussion and decision-making regarding the reconstruction plans, in particular the housing reconstruction plans. For example, during the design and planning phase, CAUPD carried out several consultations with the affected people who were going to move into New Beichuan (Zhu *et al.* 2012). Jing (2012) explained that by interviewing the affected people, the impact and capacity assessment reports could better address their needs. The contents of the reconstruction plans, compensation for relocation, housing supplies, and equality and justice were taken into consideration during this consultation procedure. It was also explained that the reconstruction planning strategies were disseminated through TV programmes and videos (Jing 2012).

Li and Lin (2013) argued that this decision-making model among the government, the public and the NGOs could vary depending on the power relationship between these three stakeholders in each community. Observing the activities post-disaster, such as search and rescue, clear up, logistics supply, assistance in housing construction, and care for children and the elderly in the communities, Luo and Fang (2013) found that the public involvement and participation could be divided into two types: self-organised and local government-organised activities. More affected people participated in the former activities (29.6 per cent) than the latter (18.5 per cent) (Luo and Fang 2013: 100). Friends and families of government

TABLE 5.1 Where the affected populations were relocated.

Relocated population	Relocated to another town		17,332
	Relocated within the same town	Relocated in different village	4,950
		Relocated within the same village	13,156
Total			35,438

Source: CAUPD 2008: 16.

officers participated in the government-led activities. Although it was easier for the government officers to mobilise these people to get involved in the recovery activities, it was not beneficial in the long term for community development. Those who participated in the self-organised activities tended to have a higher sense of identity of the community (Luo and Fang 2013). Jiang and Hou (2009) examined the affected population in Leigu Town and found that they were not enthusiastic about engaging in the decision-making process. They found the decisions and policies that they could get involved with were limited. At the same time, their impact was similarly limited in the final decision-making (Jiang and Hou 2009).

Challenges for reconstruction in Beichuan

By September 2010, Shandong Province had already sent 210 officials and 25,000 workers, provided US$2.4 billion, and completed 218 projects (Xu and Lu 2008: 82). New Beichuan (Anchuang) was established and all of the affected populations in Qushan were moved to the newly established area. Most of the infrastructure and public services were ready to use. Although the reconstruction work was one year ahead of schedule (Zhu 2011), this was not without challenges.

First of all, it was argued that the centralised approach for the one-on-one assistance programme meant that resources and personnel could be allocated and distributed to the affected areas more effectively (Hu *et al.* 2010). However, some of the assistance did not really meet local needs. Although the aid workers made efforts to speed up the reconstruction, their lack of familiarity with the local lifestyle and local culture meant they were unable to produce a consistent technical plan for reconstruction to suit the rapid change in local demand (Ge *et al.* 2010: 24). This also reflected on the appropriateness of the urban plans: although the intention to use tourism to boost the local economy was encouraging, the lack of a holistic view on the local resources, capacity and ability of the residents to operate a new industry undermined the effectiveness of the projects (Gu *et al.* 2011).

In particular, reconstruction should not only focus on the rebuilding of the infrastructure; culture and heritage are also important. Most of the affected population are members of the Qiang minority. Their society, culture and history are very different from the majority Han people. Although the Chinese government realised the importance of recognising the society, culture and history of the Qiang, and has attempted to preserve their heritage (Chen 2012), the challenge is to preserve the more intangible aspects of their culture. For example, a number of watch towers of the old Qiang tribe were restored in several locations; descriptions of their functions were illustrated on a board next to the construction for tourists. However, this was of no real use and it has lost the original meaning in the ritual parts of their culture. Also, the modern facilities provided in the new apartments, such as cookers and heating systems, are not appropriate for the Qiang's daily routine. Some residents found it too costly to use electricity, so they built shelters outside their houses to burn wood for heating and a traditional stone-based oven for cooking (Wong 2012). Although the new residential sites are comparably safer than their original habitat areas, the difference between new ways of living and the old houses of the affected population have generated new conflicts (Gu *et al.* 2011).

Since the reconstruction in Beichuan had attracted the attention and involvement of the Prime Minister, the new town was expected to become an example of best practice and also to demonstrate the characteristics of 'safety, liveability, prosperity, uniqueness, modernity, and harmony' (Zhu *et al.* 2012). These expectations have consequently placed much additional pressure on those responsible for the reconstruction work. Although it was with good intentions that the government accelerated the reconstruction work from three years to one year, much of the infrastructure is not well established, and the new buildings struggled to meet disaster risk reduction standards (Gu *et al.* 2011).

What is more fundamental is that the Chinese social security system, which only insures the urban population, has created a two-tier society (Otani 2014). Under this system, most young people working in the big cities have left older parents and young children in the countryside. Because Beichuan is located in a rural area and has an agriculture-based economy, young people are inevitably motivated to move out of the area to seek better careers. If the social security system is not changed, it will not be easy to persuade these young people to return to the area and facilitate its development.

In general, it was argued that the immediate emergency response, technical support and relief assistance were effective (Ge *et al.* 2010), but the short-term recovery, long-term reconstruction and further sustainable development lacked the incorporation of local needs (Ge *et al.* 2010). The achievements of the development plan were not as good as expected. For example, the population growth in the new town was slower than hoped. Although it aimed to reach a total population of 50,000 by 2015, the number of residents was still under 30,000 in 2014 (Naoya *et al.* 2014). The slow progress on sustainable development, the high unemployment rate, the level of poverty and the poor management of public services could slow down or even prevent urban development. Although huge amounts of finance and subsidies were invested to help with the relocated population, it was found that outside of New Beichuan, many villages were still awaiting reconstruction or had been abandoned (Naoya *et al.* 2014).

Conclusion

The number of large-scale disasters, both natural and man-made, is increasing steadily (EM-DAT 2016). Although mortality levels resulting from disasters have decreased, response, recovery and reconstruction remain major political and socio-economic challenges. The lessons of the post-Wenchuan earthquake reconstruction are also applicable to other countries and regions.

The 'one-on-one assistance programme' is claimed to be effective, despite some criticisms. It is argued that the systematic organisation and operation model facilitated effective cooperation between aid providers and receivers (Xu and Lu 2008). This operational model could be applied to similar disaster-prone cities, towns and counties. What is more beneficial is that, with the predefined partnership before a disaster hits, the assistance model could be developed in a more holistic way. In addition, through communication and understanding during normal times and situations, it would be easier for the supporting side to provide the appropriate relief, goods and services during and after a disaster.

The recognition of NGOs to provide support in disaster management recovery is a new achievement in China. 2008 was the first year of NGO involvement in China (Otani 2014). It is crucial for the government to establish relevant regulations for a normal route for NGO involvement. A clearer definition of the roles and responsibilities of the government and NGOs would enable more effective collaboration between these two sectors.

What is noted is that while most reconstruction studies focus on the rebuilt communities themselves, there is also a need to study the impact of major work or new construction projects on the surrounding neighbourhoods. There can be profound impacts on existing nearby communities in both the formal and informal social sector, taking into account the impact on the economy, society and security.

Planning for post-disaster recovery and reconstruction before a disaster occurs can effectively enhance community resilience (Johnson 2014). Recommendations in this chapter identify strategies for recovery and reconstruction that should be embedded in the post-event response phase. Since the characteristics and impact of each disaster are different, decision-makers should consider the worst-case scenario according to each type of disaster, and develop strategies to intervene appropriately in order to ensure that the impacts can be minimised.

References

CAUPD (2008) *New Beichuan Qiang Autonomous County Overall Plan*. China Academy of Urban Planning and Design. Beichuan: CAUPD and Beichuan Government (in Chinese). Available at: www.china-up.com:8080/512dz/planning/1.asp (accessed 14 April 2016).

Chen, T. (2012) The rescue, conservation, and restoration of heritage sites in the ethnic minority areas ravaged by the Wenchuan earthquake. *Frontiers of Architectural Research* 1: 77–85.

EERI (2008) *Learning from Earthquakes: The Wenchuan, Sichuan Province, China, Earthquake of May 12, 2008*. EERI Special Reports. Available at: www.eeri.org/site/images/eeri_newsletter/2008_pdf/Wenchuan_China_Recon_Rpt.pdf.

EM-DAT (2016) *Natural Disasters Trend: World 1900–2016*. Available at: www.emdat.be/disaster_trends/index.html.

Feng, S., Lu, J., Nolen, P. and Wang, L. (2016) The effect of the Wenchuan earthquake and government aid on rural households. In: Chen, K.Z., Zhang, Q. and Hsu, C. (eds) *Earthquake Lessons from China: Coping and Rebuilding Strategies*. Washington, DC:

International Food Policy Research Institute. Available at: http://ebrary.ifpri.org/cdm/ref/collection/p15738coll2/id/130269.

Ge, Y., Gu, Y. and Deng, W. (2010) Evaluating China's national post disaster plans: the 2008 Wenchuan earthquake's recovery and reconstruction planning. *International Journal of Disaster Risk Science* 1(2): 17–27.

Gu, L., Xiang, M. and Li, Y. (2015) Evaluation of six years of reconstruction since the 2008 Sichuan earthquake. Sendai Public Forum in Community-Based Reconstruction of Society and University Involvement: East Japan Lessons Compared with Kobe, Aceh, and Sichuan, Organised by Universities of Kobe, Tohoku, and Iwate, along with the 3rd World Conference on Disaster Risk Reduction, Sendai, Japan. Available at www.wcdrr.org/conference/events/339 (accessed 7 March 2016).

Hirai, K. (2009) The minutes of the Kobe-Chi-Chi-Chuetsu meeting. In Kobe, Chi-Chi, Chuetsu, Wenchuan Earthquake Recovery Conference Proceeding I-12–18 (in Japanese).

Hu, X., Salazar, M.A.M., Zhang, Q., Lu, Q. and Zhang, X. (2010) Social protection during disasters: evidence from the Wenchuan earthquake. *IDS Bulletin* 41(4): 107–115.

Huang, Y., Fu, U. and Wong, H. (2014) Challenges of social workers' involvement in the recovery of 5.12 Wenchuan earthquake in China. *International Journal of Social Welfare* 23: 139–149.

Jiang, B. and Hou, C. (2009) On the status and the mode of villagers' participation in rural community construction: a case study on the post-disaster reconstruction in Leigu, Beichuan County. *Journal of Social Work* 7: 32–35 (in Chinese).

Jing, F. (2012) Public participation in post-disaster reconstruction plan of New Beichuan Town. 48th ISOCARP Congress, Perm, Russia. Available at: www.isocarp.net/Data/case_studies/2129.pdf.

Johnson, L.A. (2014) Long-term recovery planning: the process of planning. In: Schwab, J.C. (ed.) *Planning for Post-Disaster Recovery: Next Generation*, chapter 6. PAS Report 576. American Planning Association, pp. 92–119. Available at: www.fema.gov/media-library-data/1425503479190–22edb246b925ba41104b7d38eddc207f/APA_PAS_576.pdf.

Kamata, F. (2011) *Three Years Post-Sichuan Earthquake in China: The Challenges of Reconstruction.* Tokyo: National Diet Library (in Japanese). Available at: www.ndl.go.jp/jp/diet/publication/refer/pdf/072805.pdf.

Li, S.U. and Lin, T.H. (2013) Resilience and reconstruction: the state and local community after the 2008 Sichuan earthquake. *East Asian Studies* 44(2): 1–38 (in Chinese).

Liu, Q., Ruan, X. and Shi, P. (2011) Selection of emergency shelter sites for seismic disaster in mountainous regions: lessons from the 2008 Wenchuan Ms 8.0 earthquake, China. *Journal of Asian Earth Science* 40: 926–934.

Lu, Y. and Xu, J. (2014) NGO collaboration in community post-disaster reconstruction: field research following the 2008 Wenchuan earthquake in China. *Disasters* 39(2): 258–278.

Luo, J.D. and Fang, Z. (2013) Social capital and self-organization: a study on government and self-organized activities in a post-disaster reconstruction process. *Waseda Rilas Journal* 1: 99–107.

Naoya, M., Chiho, O., Gu, L. and Kenji, O. (2014) Study on the urban and regional planning of post 2008 Sichuan earthquake restoration, China: case study on the relocation and reconstruction plan of Beichuan County Seat. *Reports of the City Planning Institute of Japan* 13 (February): 188–191. Available at: www.cpij.or.jp/com/ac/reports/13–4_188.pdf.

Otani, J. (2014) Challenges of reconstruction strategies of Chinese society post Sichuan earthquake. Special Issue: Large scale disasters and social protection I. *Overseas Social Security Research*, 187 (Summer): 4–19 (in Japanese).

PGPWERR (2008) *The State Overall Planning for Post-Wenchuan Earthquake Restoration and Reconstruction.* Planning Group of Post Wenchuan Earthquake Recovery and

Reconstruction, the Central People's Government of the People's Republic of China (in Chinese). Available at: www.gov.cn/wcdzzhhfcjghzqyjg.pdf.

Roney, B. (2011) Earthquakes and civil society: a comparative study of the response of China's nongovernment organisations to the Wenchuan earthquake. *China Information* 25(1): 83–104.

State Council (2008a) *Post Wenchuan Earthquake Reconstruction Regulations.* The Central People's Government of the People's Republic of China (in Chinese). Available at: www. gov.cn/zwgk/2008–06/09/content_1010710.htm.

State Council (2008b) *The Announcement of One on One Assistance Programme Post Wenchuan Earthquake.* The Central People's Government of the People's Republic of China (in Chinese). Available at: www.gov.cn/gongbao/content/2008/content_1025941.htm.

State Council (2008c) *Overall Plan for Post Wenchuan Earthquake Reconstruction Regulations.* The Central People's Government of the People's Republic of China (in Chinese). Available at: www.gov.cn/wcdzzhhfcjghzqyjg.pdf.

State Council (2016) *Civil Contingency Act 2016.* Available at: www.gov.cn/zhengce/con tent/2016–03/24/content_5057163.htm.

UNCRD (2009) *Report on the 2008 Great Sichuan Earthquake.* Hyogo: United Nations Centre for Regional Development Disaster Management Planning, Hyogo Office. Available at: www.recoveryplatform.org/assets/publication/UNCRD_Sichuan_Report_ 200903EN.pdf.

USGS (2008) *M7.9 Eastern Sichuan, China, Earthquake of 12 May 2008 and Aftershocks through 28 May 2008.* United States Geological Survey. Available at: https://commons.wikimedia org/wiki/File:2008_Sichuan_Earthquake_aftershockes_through_May_28.pdf.

Wei, Y.-M., Jin, J.-L and Wang, Q. (2012) Impact of natural disasters and disasters risk management in China: the case of China's experience in Wenchuan earthquake. In: Sawada, Y. and Oum, S. (eds) *Economic and Welfare Impacts of Disasters in East Asia and Policy Response.* ERIA Research Project Report 2011–8. Jakarta: ERIA, pp. 641–675. Available at: www.eria.org/Chapter_17.pdf.

Wong, H. (2012) A reflection on community humanity rebuilding after the Sichuan 5.21 earthquake. In: Yang, Z.H. and Yang, Z.P. (eds) *Green Start: Re-examination of the relation between Nature and Humanity.* Taipei: Yuan-Liou Publishing Co. Ltd (in Chinese).

World Economic Forum (2010) *Engineering and Construction Disaster Resource Partnership: A New Private–Public Partnership Model for Disaster Response.* Geneva: WEF. Available at: www3.weforum.org/docs/WEF_EN_DisasterResourcePartnership_Report_2010.pdf.

Xu, J. and Lu, Y. (2008) A comparative study on the national counterpart aid model for post-disaster recovery and reconstruction: 2008 Wenchuan earthquake as a case. *Disaster Prevention and Management* 22(1): 75–93.

Zhang, Q. and Yu, X. (2009) *Research on NGO Participation in Reconstruction Following the Wenchuan Earthquake.* Beijing: Peking University Press (in Chinese).

Zhu, J.X. (2011) Beichuan, Shangdong supports you. In: Zhu, D.F. and A. L. (eds) *National Action: The Support from the 19 Provinces/Cities in the National One-on-One Programme.* Chengdu: Sichuan Literature and Art Publishing House (in Chinese).

Zhu, Z., Li, M. and Huang, S. (2012) Post-quake reconstruction planning and implementation for Beichuan New Town. 48th ISOCARP Congress, Perm, Russia. Available at: www.isocarp.net/Data/case_studies/2149.pdf.

6

PARTICIPATION FOR DISASTER RESILIENCE

A life cycle approach to reconstruction projects in India

Mittul Vahanvati

Introduction

India is one of the top ten countries in the world at risk from natural disasters (NIDM, 2001). The toll of disasters on human lives and the national economy has risen substantially since 1999, with total losses of approximately 2 per cent of national GDP, annually (GoI-UNDP, 2011; Guha-Sapir *et al.*, 2012). Vulnerability of the built environment is seen as one of the major reasons for India's disaster risks (UNNATI *et al.*, 2012). However, underlying socio-cultural and financial issues are often the root cause of the built environment's vulnerability. In a post-disaster context, scholars warn that addressing technical resilience without incorporating strategic issues, such as social or economic resilience, may hardly empower the communities concerned (Davis, 1978; Berke *et al.*, 1993; UN-Habitat, 2012; Ahmed, 2011; Lyons *et al.*, 2010; Mulligan and Nadarajah, 2012). To this end, an international consensus has developed for participation— as a means as well as an end—for enabling disaster resilience of communities (UNDRO, 1982).

India being a federation, the states have the primary responsibility for disaster recovery management. For the first time in the history of India, reconstruction after the 2001 Gujarat earthquake saw an 'Owner Driven Reconstruction' (ODR) approach (NDMA, 2005). Since 2001, the Indian national and state governments have progressively amended their reconstruction policies in order to shift from a top-down or a relief-based approach to a participatory/empowering approach. An ODR approach was also adopted in 2005 following the Kashmir earthquake and in 2008 following the Kosi River flooding in Bihar (Barenstein and Iyengar, 2010). Despite clear policy commitments to maximise Owner Driven Reconstruction, its practice continues to remain patchy and sporadic, with undesirable long-term implications.

In order to determine how a participatory reconstruction project after disaster can enhance the long-term disaster resilience of communities at risk, the researcher asks the following question:

• What approaches to community participation during ODR are most likely to enhance community confidence, awareness and livelihoods in order to maintain their houses' and their settlement's safety and for their wider disaster resilience in the long term?

Methodology

Four good practice reconstruction projects were selected from India for case study investigation. Two projects are from the state of Gujarat following the 2001 earthquake—Hodko settlement and Patanka settlement. The other two are from the state of Bihar following the 2008 Kosi River floods—Orlaha and Puraini settlements (see Table 6.1).

The primary reason for the selection of these four case studies was that the Civil Society Organisations (CSOs) had an upfront responsibility for capacity building and going beyond the rebuilding of resilient houses. Another reason was that both CSOs were involved in Gujarat as well as Bihar. The two CSOs were:

i)   Kachchh Nav Nirman Abhiyan, hereafter referred to as Abhiyan.
ii) The Sustainable Environment and Ecological Development Society (SEEDS).

Hence, it was expected that these two CSOs would have developed some insights on participatory approaches. They would know what worked, what did not, why and, most importantly, how ODR approaches evolved over the course of seven years (2001–2008).

TABLE 6.1 Criteria for the selection of case studies.

	Reconstruction programme	Post-2001 Gujarat earthquake		Post-2008 Bihar Kosi River flooding	
	Agency	CS-1 Abhiyan Hodko	CS-2 SEEDS Patanka	CS-3 ODRC Orlaha	CS-4 ODRC Puraini
Key selection criteria	Owner-driven/ participatory	√	√	√	√
	Resilience features explicitly incorporated in housing	√	√	√	√
	Upfront responsibility for skills training and livelihood consideration	√	√	√	√

INCREASING IMPACT ON THE DECISION

	INFORM	CONSULT	INVOLVE	COLLABORATE	EMPOWER
PUBLIC PARTICIPATION GOAL	To provide the public with balanced and objective information to assist them in understanding the problem, alternatives, opportunities and/or solutions.	To obtain public feedback on analysis, alternatives and/or decisions.	To work directly with the public throughout the process to ensure that public concerns and aspirations are consistently understood and considered.	To partner with the public in each aspect of the decision including the development of alternatives and the identification of the preferred solution.	To place final decision making in the hands of the public.
PROMISE TO THE PUBLIC	We will keep you informed.	We will keep you informed, listen to and acknowledge concerns and aspirations, and provide feedback on how public input influenced the decision. We will seek your feedback on drafts and proposals.	We will work with you to ensure that your concerns and aspirations are directly reflected in the alternatives developed and provide feedback on how public input influenced the decision.	We will work together with you to formulate solutions and incorporate your advice and recommendations into the decisions to the maximum extent possible.	We will implement what you decide.

FIGURE 6.1 Public participation spectrum.
Source: International Association for Public Participation.

The case studies are divided into four phases based on a project life cycle approach for logical framework analysis (LFA), as follows:

- Phase-I: Planning and design input
- Phase-II: Construction output
- Phase-III: Project results in the short term
- Phase-IV: Long-term impact.

Participation in each phase is rated using the Spectrum of Participation (IAP2) (see Figure 6.1). Within each case study, data was purposely collected from beneficiaries, non-beneficiaries and CSO members to ensure validity. Within these case studies, social profiling was used to identify beneficiaries for interviews. This allowed the author to understand the influence of caste and power hierarchies in recovery assistance. Semi-structured interviews, focus group discussions, photographs and sketches of houses and CSO publications have informed the discussion. The author conducted field studies in 2012 and 2014. To ensure the confidentiality of research participants, identity codes are used, such as HA-X 2014, where 'H' refers to the location, 'A' refers to agency member and 'X' is the number assigned to each respondent followed by the year of interview.

The 2001 Gujarat earthquake

Earthquake impact and reconstruction policy

On 26 January 2001 (Indian Republic Day), the western state of Gujarat was hit by a massive earthquake measuring 7.9 on the Richter scale. This was the second largest and the deadliest earthquake in Indian history (GoI-UNDP, 2011). The earthquake caused nearly 20,000 deaths (UNDP, 2009) and destroyed over 1 million houses (GoI-UNDP, 2011) (see Figure 6.2(a)). Kachchh district of Gujarat was the epicentre of this earthquake. When the earthquake hit, Kachchh had not yet fully recovered from over three years of drought (UNDP, 2009).

Within a week of the earthquake the government of Gujarat had set up a nodal agency to manage the disaster recovery, named the Gujarat State Disaster Management Authority (GSDMA). Soon after, the government announced a reconstruction policy, whose key features were:

i)   An Owner Driven Reconstruction (ODR) with funding assistance from the World Bank.
ii) Public–private partnerships for implementing reconstruction (UNDP and Abhiyan, 2005).

Case study 1

Hodko settlement in the Kachchh district was very close to the epicentre of the earthquake and suffered major destruction. Hodko settlement is located in a hot, arid zone, with desert and grasslands (termed a *Banni* region). The traditional houses in Hodko (called *bhungas*) were built using mud and grass (see Figure 6.2(b)). Hodko continues to be known for its colourful traditional crafts (embroidery, leatherwork, etc.), dairy farming and animal husbandry. The region faces severe droughts, moderate cyclonic winds, earthquakes and occasional flash floods.

Hodko has two predominant castes—i) *Haleputra* (the royal Muslims, originally from the Sindh region of Pakistan) and ii) *Meghwal* (Harijan Hindus or the untouchables). Despite the evident political tensions between the Hindus and Muslims in other parts of India, the two communities lived harmoniously in Hodko (see Table 6.2).

TABLE 6.2 Social profile of Hodko, Kachchh district in Gujarat.

Status (high to low)	Social profile of Hodko	Livelihood
1	Muslim castes (about18 different castes)—Haleputra predominant in Hodka	Animal husbandry—they are a pastoral community (also termed Maldhari) who breed buffaloes, cows, goats, camels and provide skins to the Hindus for processing.
2	Hindu castes—Meghwal and Vadha Koli (Marwada Dalits or the untouchables)	Service providers—convert skins of cattle provided by Muslims into leather products and do wood carving, whereas women do embroidery.

Planning and design

Abhiyan is a locally based network of 26 organisations (CSOs) that were work-ing with the community prior to the earthquake. Thus, Hodko residents trusted and opted to work with Abhiyan. Abhiyan set up informal shelter cluster hubs, which were later known as *Setu Kendras* (literally meaning bridging centres). The community was an integral part of each *Setu*. This included two village motiva-tors (*gram preraks*), who worked with five professional staff from Abhiyan—social workers, information manager, accounts officer and local civil engineer (UNDP, 2001). The community was able to channel its concerns to the government via the *Setu* in a collective and effective manner. Some of the victories that *Setus* had in influencing the government/GSDMA were: i) reassessment of housing dam-age categories; ii) amendment of policy from relocation to *in-situ* reconstruction; and iii) an agreement on a community-led or an ODR approach for sustainable rehabilitation (UNDP, 2001). This initial exercise established and strengthened a relationship of trust between the Hodko community and Abhiyan.

At the settlement scale, Abhiyan put together an assistance package for recon-struction. As per the package, Abhiyan would provide a core shelter design (18 square metres), all construction materials and labour, while the residents had to contribute a minimum of 10 per cent of each house's cost—in the form of either labour or money (HA-1 2014). Abhiyan also legalised the mud technol-ogy (HA-2 2014; KMVS, 2001). Fifty-six out of ninety-seven Harijan families signed an agreement to participate in Abhiyan reconstruction (KMVS, 2001). The Muslims declined any assistance as they were more affluent (HA-2 2014). The settlement layout was drafted by the residents while retaining existing plots, based on traditional laws. The design, being a core shelter, did not allow for modifications by Hodko residents. Despite the generally progressive nature of Gujarat society, it was observed that women's participation was a lot less than men's. Overall, community control of various aspects of the initial planning and design phase is rated as 'collaborative'.

Construction

As a first step before construction, Abhiyan built model houses for the neediest residents in the settlement (widows, the elderly and disabled households) (Desai, 2002; Gupta and Shaw, 2003). During this process, locals were also trained in safe construction skills. Due to the scale of the construction work, relatively more non-local masons were trained to build resilient shelters (HA-3 2014). While most local residents 'had good friendship and skills in mud' (HA-1 2014), they lacked skills in safe construction. Construction monitoring was entirely managed by Abhiyan, while material purchase and distribution were managed collabora-tively (HA-4 2014). Thus, participation during the construction phase is rated as 'involved'.

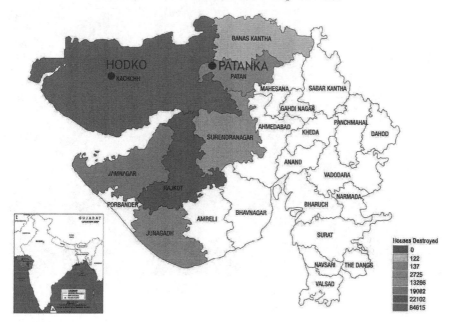

Earthquake-affected districts of Gujarat State

Houses Destroyed	
	0
	122
	137
	2725
	13266
	19082
	22102
	84615

FIGURE 6.2 (a) Map of Gujarat with two case-study settlements, Hodko and Patanka. *Source*: UNDP, 2001.

Hodko traditional village

FIGURE 6.2 (b) Hodko traditional village.

FIGURE 6.2 (c) Patanka earthquake damage.
Source: Photo by Rameshbhai Thakor, SEEDS.

Short-term outcomes

In 15 months, 56 families in Hodko had completed the rebuilding of resilient houses (two *bhungas* per family) (see Figure 6.3). The satisfaction among all the residents interviewed was very high in terms of participatory process and appropriateness of house design, construction quality, disaster safety, cost effectiveness and low maintenance. However, there was a bit of discontent among some residents regarding the climate comfort of the houses and among some artisans regarding fewer locals being trained (HA-7 2014). Abhiyan's work was recognised as best practice in reconstruction (UN-Habitat *et al.*, 2008; UN-Habitat *et al.*, 2009).

Abhiyan did not stop there. It took the momentum from housing reconstruction further by moving onto other livelihood enhancement and community empowerment projects (HA-2 2014).

Long-term impact

Fifteen years later, almost all the residents in Hodko had personalised their rebuilt houses and continue to reside in them as they feel confident in their robustness. However, the residents' memory of the resilient features in the house design was fading. Additionally, it was not easy to access the resources (shuttering, brick press machines or skilled masons) required to build seismic-safe house extensions. This was evident in the new houses or extensions built that are incrementally discontinuing the use of proposed materials and technologies essential for housing resilience.

FIGURES 6.3 (a–c) Rebuilt houses (past and present) in Hodko settlement.

On the other hand, the *Sham-e-Sarhad* eco-resort project has been highly successful. This project has increased the financial capacity, cultural pride and social well-being of Hodko residents, who no longer identify themselves as poor or backward (HA-2 2014). Moreover, the resort directs some of the profit towards village infrastructure and amenity improvements. Despite an increase in livelihood, however, the residents of Hodko are not currently investing in resilient housing.

(d)   **(e)**

(f)

FIGURES 6.3 (d–f) Rebuilt houses (past and present) in Patanka settlement.

Case study 2

Patanka settlement in Patan district was located further from the epicentre of the earthquake, but it also suffered high damage. The climate of Patanka is similar to that in Hodko, but traditional houses were built from yellow sandstone with mud plaster (see Figure 6.3(d)). It is well known for its cumin farming. Contrary to Hodko, Patanka settlement had only one predominant caste—*Aahir* (Brahmins—the highest Hindu caste)—with small representation of others (see Table 6.3) (GoB and ODRC, 2008).

TABLE 6.3 Social profile of Patanka, Patan district in Gujarat.

Status (high to low)	Social profile of Patanka, Patan, Gujarat	Livelihood
1	Aahirs (70%) (Brahmins)	Farming, animal husbandry (well educated)
2	Koli Thakurs (16%) (Rajputs)	Farming, animal husbandry (moderately educated)
3	Rabaris (10%)	Pastoralism and labour (not educated)
4	Harijans (4%), few Suthars (carpenters), Bawajis (monks) and one Nai (hairdresser)	Farming labour, animal husbandry

Planning and design

Five months after the earthquake, SEEDS approached the community of Patanka was to offer assistance. Unlike Abhiyan, SEEDS was neither a local non-government organisation (NGO) and nor did it have a relationship with this particular community. SEEDS' efforts at forming a partnership with the local NGO—the Self-Employed Women's Association (SEWA)—proved futile. SEEDS organised 'shake-table' tests to showcase seismic-safe technologies among the local residents (Gupta and Shaw, 2003). This exercise earned SEEDS the trust of local residents. With mutual consent, a tripartite agreement was made between GSDMA, SEEDS and the Patanka village *Panchayat* (local elected members). As part of this agreement, the residents of Patanka would receive GSDMA funding with a top-up from SEEDS (in the form of steel and cement for construction). All 225 families in Patanka decided to accept SEEDS' assistance (PA-1 2014). SEEDS gave Patanka residents the freedom to design their own house core of about 12 square metres. The only constraint was to abide by the multi-hazard safety features in housing. Overall, community participation during the initial planning and design phase is rated as 'involved'.

Construction

Like Abhiyan, SEEDS also constructed a model house for a widow and trained local masons in safe construction skills as their construction quality was very poor (PA-4 2014) (Vahanvati and Beza, 2016). Additionally, two highly skilled masons were invited by SEEDS from the National Society for Earthquake Technology in Nepal to live in Patanka for three months and train the locals (PA-4 2014). The residents procured all their construction materials (except for steel and cement), employed labour and monitored the construction quality collaboratively (PA-2 2014). The community participation during the construction phase is rated as 'collaborative'.

Short-term outcomes

Three hundred families managed to build resilient houses with support from SEEDS (see Figure 6.3). The satisfaction among Patanka residents was as high as in Hodko. SEEDS also continued onto other projects, such as building a school and a water tank to cope with future droughts. Most importantly, SEEDS mobilised a guild of 15–20 trained masons from Patanka to form a SEEDS Mason's Association (SMA) (PA-4 2014). SEEDS' work in Patanka was recognised as best practice in going beyond rebuilding of houses in order to enhance 'community resilience' (IFRC, 2004).

Long-term impact

Fifteen years later, satisfaction and confidence were as high in Patanka as in Hodko; however, the memory of the resilient features was much better. Livelihood of only the trained masons who joined SMA had increased. The reason was that the SMA masons were supported by SEEDS for over nine years with further training, government certification in safe construction skills and employment (PA-2, PA-4 2014). In the long term, despite high awareness and an ease of access to skilled masons, only half of the residents were investing in their houses' continuing safety.

The 2008 Bihar Kosi floods

Flood impact and reconstruction policy

In August 2008, the north Indian state of Bihar was devastated by a rupture in the embankment of the Kosi River. There was a sudden surge and a change in the natural course of the Kosi River, inundating the so-called 'protected areas' that had not experienced flooding for several decades (GoB, 2010; UNDP, 2009). The scale of damage was exceptionally high. Over 3 million people were affected (PiC, 2010; UNDP, 2009), more than 200,000 homes were damaged and a large numbers of cattle and crops were impacted due to protracted inundations (FMIS and GoB, 2009).

Within four months of the disaster (in December 2008), the government of Bihar announced an Owner Driven Reconstruction (ODR) policy. The government assembled a network of institutions and agencies from all over Asia, termed the Owner Driven Reconstruction Collaborative (ODRC) (GoB and ODRC, 2008). The ODRC comprised UNDP, Indian national and Gujarat government authorities, Abhiyan, the Asian Coalition for Housing Rights (ACHR), SEEDS and the World Habitat Centre, among others (GoB and the ODRC, 2008). The role of the ODRC was to assist the government in policy formation and implementation. Some unprecedented aspects of this reconstruction programme were:

i) Piloting before ODR policy announcement for contextualising the policy.
ii) Involvement of the ODRC from the initial policy formation stages.
iii) The state taking responsibility for ODR implementation.

The two ODR pilot settlements of Orlaha and Puraini have been selected as case studies.

Case study 3

Orlaha settlement was situated far from the burst Kosi River embankment (see Figure 6.4). Both Orlaha and Puraini have humid sub-tropical climates and are located in the far north of Bihar, also known as the Kosi River basin. This region is characterised by a network of rivers, land shortages, poverty and illiteracy with limited access to electricity. The region has an abundance of bamboo, grass and mud-based items such as bricks and roofing tiles. The Kosi River basin is prone to annual flooding, earthquakes and cyclonic winds.

In Bihar's Kosi region, residents identify their community based on their religion, caste or lineage. There are four communities who coexist harmoniously despite the social hierarchy of castes and practice barter for sustenance (see Table 6.4). The ODRC chose pilot villages to demonstrate an ODR process and to build the government's capacity in managing disaster recovery. Since the ODRC was a non-local consortium of national and international organisations, it partnered with a local CSO, Gramsheel.

Planning and design

Four months after the floods, an agreement was made between the residents of Orlaha and Puraini, the ODRC and the government of Bihar participate in an ODR pilot project. As per the agreement, the government identified the beneficiaries

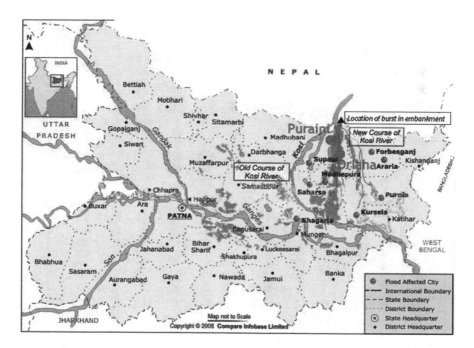

FIGURE 6.4 (a) Map of Bihar with two case-study settlements, Puraini and Orlaha.
Source: www.mapsofindia.com (copyright permission granted).

FIGURE 6.4 (b) Flood devastation.
Source: GoB, et al. (2010). Bihar Kosi Flood (2008) Needs Assessment Report, World Bank, p. 13.

Traditional house in Bihar Kosi river basin

FIGURE 6.4 (c) Traditional house in Bihar Kosi River basin.
Source: GoB, et al. (2010). Bihar Kosi Flood (2008) Needs Assessment Report, World Bank, p. 13.

TABLE 6.4 Social profile of Orlaha and Puraini settlements in Bihar.

Social profile of Orlaha	Status (high to low)	Social profile of Puraini
Mandal (agricultural land owners, contractors)	1	Mandal (agricultural land owners/master masons) (90%)
Patwa (labourer—agricultural/ construction)	2	Rajput (driver, migrant labourer)
Muslim (labourer—cotton quilt makers)	3	Mehta (land owners)
Sardar (labourer—masons/ bamboo artisans)	4	Harijan/Musahars (labourers)

and the ODRC implemented the project. To maintain transparency in communication, *Kosi Setu Kendras* (KSKs) were set up, as in Gujarat. The ODRC aimed at instilling faith in the local residents to be able to rebuild their resilient houses (GoB, 2010). To do so, the ODRC provided socio-technical support, legalised traditional bamboo technology and published a handbook with a few core shelter designs and the non-negotiable resilience features, all in the local language (Hindi) (BIPARD and The Shelter Group, 2008).

With the ODRC's support, the local residents resolved land issues of the landless, planned their settlement layout and designed their own houses. The ODRC facilitated the opening of a bank account for each family (in male and female names). In a region where there is high illiteracy and poverty, gaining community trust in opening a bank account was the most challenging task. Overall, community participation in this initial phase in both Orlaha and Puraini is rated as 'empowering'.

Construction

Similar to Gujarat, the ODRC built model houses to demonstrate a palette of technology options. They also trained residents in safe construction skills. Most residents had good construction skills but lacked finesse and knowledge of safe construction (BA-4 2014). Engineers were hard to find in the villages, so a small number of trained masons and *dabia mistries* (bamboo artisans) were employed as *rajmistries* (master masons) in place of engineers, to be part of the ODRC technical team (BA-1 2014). The KSK configuration was similar to that in Hodko: one *rajmistry*, one engineer, one manager and two social workers. Each KSK team was responsible for providing day-to-day handholding support to the residents of one village or approximately 200 houses.

When compared to Gujarat, key differences with regards to community participation in Bihar are as follows:

i)   Residents were enabled to make decisions for technology selection, material procurement and labour selection.
ii) A palette of contextually appropriate construction technologies was offered.
iii) Training in safe construction was provided to the local residents.

iv) Funding was allocated for basic amenities, landscaping, infrastructure and loss of livelihood due to engagement in construction, apart from housing reconstruction.

v) Time-based incentives were offered for speedy construction (six months).

Community participation during construction in Orlaha and Puraini is rated as 'empowering'.

Short-term outcomes

In six months, 41 families had rebuilt resilient houses in Orlaha settlement using bamboo construction technology (see Figure 6.5). Almost all the residents interviewed were highly satisfied with the participatory process and the housing outcome (quality, disaster safety, cost effectiveness and low maintenance). There was a bit of dissatisfaction among residents in both villages regarding beneficiary

(a)   (b)

(c)

FIGURE 6.5 (a–c) Rebuilt houses (past and present) in Orlaha settlement.
Source: Hunnarshala.

FIGURE 6.5 (d–f) Rebuilt houses (past and present) in Puraini settlement.
Source: Hunnarshala.

selections, which was done by the government. One social worker stated that 'the assistance did not reach those who were in real need of a house at that time' (BA-3 2014). Despite the ODRC's efforts to maintain transparency and negotiate a resolution with the government, no amendments were made to the beneficiary list (BA-2 2014). In both settlements, residents stressed that without the presence of the ODRC—a non-local, non-corrupt CSO—they could not have achieved speedy and quality housing recovery—at least not with the assistance of the government alone.

Long-term impact

Eight years after the floods, the housing reconstruction programme was still ongoing. By 2015 it was entirely managed by the state government and the World Bank. The ODRC withdrew its support after the pilot village construction. In the pilot settlements, almost all the residents had personalised their houses and continued to reside in them. Some of the residents in Orlaha were questioning the longevity of their bamboo houses as the structural bamboo poles were already infected by borers. A majority of new houses and extensions were made using brick and RCC technology (not bamboo), which incorporated resilient features. Hence, at least one construction technology had found continuity.

Sadly, a majority of the waterless toilets constructed using bamboo technology were in a dilapidated condition and not in use by 2015. These waterless toilets were being replaced by regular flush toilets as residents' finances permitted. Most men and a few women had vivid memories of the disaster resilience features in their houses. There was no evidence of livelihood improvement in Orlaha result-ing from the safe construction training. In Orlaha, there was resentment among some trained *dabia mistries* regarding a lack of continued support in terms of certi-fication in safe construction skills (BNB-1 2014).

Case study 4

Except for the fact that Puraini settlement was very close to the burst embankment, nothing was different in regards to the ODR implementation when compared to Orlaha settlement (see Figure 6.4). This section only explains key issues that were different in Puraini.

In the short term, in contrast to Orlaha, almost all of the 89 houses in Puraini were rebuilt using brick and RCC (see Figure 6.5).

In the long term, contrary to Orlaha, livelihoods had evidently improved in Puraini due to training in safe construction skills. Some residents became entre-preneurs and started their own building contracting company, which employed other locals as labourers (BPB-8 2014). After the floods, Puraini residents had lost their main source of livelihood—agriculture—due to over three metres of sand deposition on land. They had no choice but to diversify their livelihoods. These residents organised and mobilised themselves, without any continued support from CSOs, to turn adversity into livelihood opportunity. Increase in livelihoods has also increased the residents' confidence.

Overall, the communities of Orlaha and Puraini have emerged more resilient than before—with robust houses, basic amenities, increased awareness and more confidence. Livelihood has increased in Puraini more than in Orlaha. Almost all the residents are investing in their families' development—for instance, in children's education and resilient housing.

Key factors for success of ODR in the long term

Three lessons have emerged from the examination of four good-practice ODR case-study projects in India:

1) Gaining community trust and local partnership—a foundation for ODR.
2) Artisanal skills training or capacity building—during ODR.
3) Continued support for enhanced community self-sufficiency—post-ODR.

Based on the findings from empirical research, a new framework for operation-alising community participation during reconstruction is derived (see Figure 6.6). The framework considers strategic issues that go beyond one project life cycle understanding; to transition from the enhancement of trust and shared aims (social resilience), to housing reconstruction (technical/physical resilience), to capacity building and the diversification of livelihoods (financial resilience). The framework is designed for use by CSOs/implementers to enhance community engagement and disaster resilience in the long term.

Gaining community trust and local partnerships: A foundation for ODR

For the survivors of a disaster, trusting an unknown CSO for housing assistance seems like a big ask, especially when they are still recovering from the trauma and

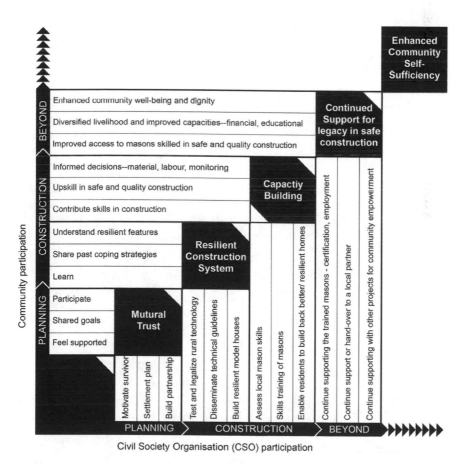

FIGURE 6.6 A new framework for operationalising community participation for disaster resilience during post-disaster reconstruction.

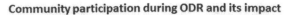
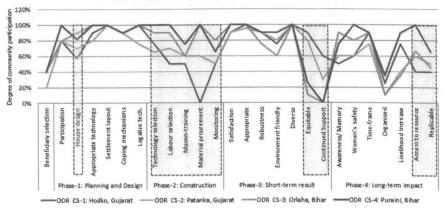

FIGURE 6.7 Degree of community participation during and after housing reconstruction process, based on quantitative data analysis (highlighted areas show major differences between the four case studies).

TABLE 6.5 Community participation rating and its long-term impact.

Case-study ODR project settlements		Participation rating			
		Planning Phase-I	Construction Phase-II	Continued support Phase-III	Long-term impact Phase-IV
CS-1	Hodko, Gujarat	• • • •	• • •	✓	Increase in livelihood
CS-2	Patanka, Gujarat	• • •	• • • •	✓	Moderate increase in livelihood and legacy of resilient housing
CS-3	Orlaha, Bihar	• • • • •	• • • • •	X	Legacy of resilient housing and disaster risk reduction
CS-4	Puraini, Bihar	• • • • •	• • • • •	X	Increase in livelihood; legacy of resilient housing and disaster risk reduction

(• • • = Involve; • • • • = Collaborate; • • • • • = Empower on IAP2 community participation spectrum)

loss of livelihood. The *Setu Kendras* in Hodko comprised one such effort to address this issue. Due to their success in Gujarat, *Setus* were also adopted in Bihar, but with modifications, such as:

- From an informal grassroots setup to a formalised setup.
- From small-scale setup in Kachchh district to a large-scale setup affecting the entire Kosi region.
- Assisted in coordination and partnerships between stakeholders; for example, the ODRC's partnership with the local CSO Gramsheel, whose knowledge of local language and cultural norms played a crucial role in gaining community trust.

On comparing Gujarat and Bihar's *Setu* setup, the formal setup seems to be a shift in the right direction—that is, towards decentralisation.

Although the *Setu* approach in India is unique, it is comparable to the shelter cluster, which assists faster and better recovery (Shelter Cluster, 2015). *Setus* have proven to be a successful mechanism at multiple levels—gaining community trust and establishing partnerships—which has laid the foundations for a successful ODR project.

Artisanal skills training or capacity building, during ODR

The case studies demonstrate that when skills training, employment and capacity building of 'locals' were given precedence over construction speed, the long-term impacts are positive. This is evident in the changes made by Abhiyan and SEEDS from 2001 to 2008. The results of community engagement during initial planning and construction activity phases, and their direct or indirect long-term impacts, are highlighted in Figure 6.7 and represented in Table 6.5. For instance, compared to Hodko (CS-1), more locals were trained during the construction phase and employed longer term in Patanka (CS-2), resulting in resilient technology being embedded in the local culture. Moreover, the increase in choices provided to local residents—in terms of design, material procurement and technology selection (in CS-3 and CS-4, unlike CS-1)—can be linked to an increase in residents' awareness and access to resources in the longer term. Overall, high resident engagement in Bihar through all phases has shown positive long-term impacts—in increased livelihood (despite lack of continued support) and regaining lost faith in residents' own ability to rebuild and maintain resilient houses.

Continued support for masons and community until autonomous, post-ODR

Gujarat's Hodko and Patanka settlements received continued support from CSOs long after the reconstruction had been completed (see Table 6.5). In contrast, the Puraini and Orlaha communities did not receive any extended support after the reconstruction had concluded. A number of lessons can be learnt from the Gujarat and Bihar case studies.

Long-term thinking, beyond one project life cycle

This was evident in the approach of Abhiyan and SEEDS during reconstruction in Gujarat. Both CSOs spring-boarded onto other developmental projects to address underlying issues, such as livelihood and drought. For example, Abhiyan livelihood projects were successful at increasing and diversifying the local livelihood, but unsuccessful at cultivating a culture of safe construction for future disaster risk reduction (DRR). There was a missing link, the answer for which was found in Patanka. In Patanka, SEEDS provided continued support to trained masons by linking their safe construction skills to income earning opportunities. These trained masons have ensured the uptake of proposed technology and are leaving a legacy of resilient construction.

Context-specific time-frame for participation

In Bihar, neither the trained masons nor the residents were provided with extended support for livelihood or certification of skills. Despite this, some entrepreneurial residents in Puraini have managed to find livelihoods from their newly acquired skills. These residents are having a catalyst effect on the livelihood of the entire Puraini community. Although a similar effect is not witnessed in Orlaha, it is worth nothing that what has happened in Puraini is very context-specific. First, since the residents of Bihar had never previously witnessed developmental assistance from CSOs they were very receptive (HA-2 2014); and, second, the residents had a sense of urgency to diversify their livelihoods after all the traditional options had been lost.

The four case studies signify that not only participation but trust building, capacity building/skills training and some extended support to trained masons plays a significant role in the uptake of proposed technology and future disaster resilience of communities. Additionally, Puraini exemplifies that the one-size-fits-all approach to the time-frame for reconstruction and continued support varies according to the particular context.

Conclusion

The focus of this chapter was to identify approaches to community participation during reconstruction projects that are most likely to enhance the disaster resilience of housing and residents in the long term. Four 'good practice' case-study reconstruction projects from the Indian states of Gujarat and Bihar were compared to identify the long-term impact of varying participatory approaches. Three key lessons were discussed and proposed in the form of a new operational framework to operationalise community participation. The first finding was the significance of social process, such as grassroots motivation of survivors during the initial planning phase, even before the beginning of construction. The second finding, about building the capacity of the local community, is nothing new for the practitioners in disaster recovery, yet it is often compromised. The third finding was about planning beyond one project life cycle—that is, beyond the rebuilding of houses—so as to ensure self-reliance of the community in terms of livelihood, awareness and safe construction skills. Based on these findings, a new framework for operationalising community participation during recovery projects is proposed. This chapter contributes this new framework, which could potentially help CSOs, donors and government authorities to understand context specificities of communities, participation, project life cycle and associated time-frame.

References

Ahmed, I. 2011. An overview of post-disaster permanent housing reconstruction in developing countries. *International Journal of Disaster Resilience in the Built Environment*, 2, 148–164.

Barenstein, J.D. and Iyengar, S. 2010. India: From a culture of housing to a philosophy of reconstruction. In: Lyons, M., Schilderman, T. and Boano, C. (eds) *Building Back Better: Delivering People-Centred Housing Reconstruction at Scale*. Warwickshire: Practical Action Publishing.

Berke, P.R., Kartez, J. and Wenger, D. 1993. Recovery after disaster: Achieving sustainable development, mitigation and equity. *Disasters*, 17, 93–109.

BIPARD and The Shelter Group. 2008. *A Design Handbook for Bamboo Houses*. Patna: Bihar Institute of Public Administration and Rural Development.

Davis, I. 1978. *Shelter after Disaster*. Oxford: Oxford Polytechnic Press.

Desai, R. 2002. *UN/Desa Project—Field Shake Table Program: For Confidence Building in Quake Resistant Building Technology*. India: National Centre for People's Action in Disaster Preparedness (NCPDP).

FMIS and GoB. 2009. *Flood Report 2009*. Patna: Flood Management Information System Cell (FMIS) and Water Resource Department, Government of Bihar.

GoB. 2010. *Reconstruction of Multi-Hazard Resistant Houses for the 2008 Kosi Flood Affected Districts in Bihar*. Patna: Government of Bihar, Department of Planning and Development.

GoB and ODRC. 2008. *Workshop on Owner Driven Reconstruction and Rehabilitation for Kosi Flood Affected Regions*. Patna: Government of Bihar, Department of Planning and Development, Owner Driven Reconstruction Collaborative.

GoI-UNDP. 2011. *Disaster Management in India*. New Delhi: Ministry of Home Affairs, Government of India.

Guha-Sapir, D., Vos, F., Below, R. and Ponserre, S. 2012. *Annual Disaster Statistical Review 2011: The Numbers and Trends*. Brussels: EM-DAT International Disaster Database, Centre for Research on the Epidemiology of Disasters.

Gupta, M. and Shaw, R. 2003. *PNY Report. Patanka Navjivan Yojna: Towards Sustainable Community Recovery*. India: Earthquake Disaster Mitigation and Research Centre (EDM); United Nations Centre for Regional Development (UNCRD); NGOs-Kobe; National Center for People's Action for Disaster Preparedness (NCDPD); Sustainable Environment and Ecological Development Society (SEEDS).

IFRC. 2004. *World Disasters Report 2004—From Risk to Resilience—Helping Communities Cope with Crisis*. Geneva: International Federation of Red Cross and Red Crescent Societies (IFRC).

KMVS. 2001. *List of Beneficiaries*. Kutch Mahila Vikas Sangathan.

Lyons, M., Schilderman, T. and Boano, C. 2010. *Building Back Better: Delivering People-Centred Housing Reconstruction at Scale,* Warwickshire: Practical Action Publishing.

Mulligan, M. and Nadarajah, Y. 2012. *Rebuilding Communities in the Wake of Disaster: Social Recovery in Sri Lanka and India*. New Delhi: Routledge.

NDMA. 2005. *The Disaster Management Act 2005*. New Delhi: Ministry of Law and Justice.

NIDM. 2001. *The Report of High Powered Committee on Disaster Management*. New Delhi: Nation Centre for Disaster Management, Indian Institute of Public Administration, Department of Agriculture and Cooperation, Ministry of Agriculture.

PIC. 2010. *Kosi Reconstruction and Rehabilitation Project*. India: PIC Consulting, Owner Driven Reconstruction Collaborative.

Shelter Cluster. 2015. About us. Available at: www.sheltercluster.org/working-group/about-us (accessed 25 July 2015).

UN-HABITAT. 2012. *Our Work: Risk and Disaster Management: Relief to Development*. Available at: www.unhabitat.org/content.asp?typeid=19&catid=286&cid=869 (accessed 26 July 2012).

UN-HABITAT, UNHCR and IFRC. 2008. *Shelter Projects*. 2008 edn. Fukuoka and Geneva: UN-Habitat, UNHCR and The International Federation of Red Cross and Red Crescent Societies.

UN-HABITAT, UNHCR and IFRC. 2009. *Shelter Projects*. 2009 edn. Fukuoka and Geneva: UN-Habitat, UNHCR and The International Federation of Red Cross and Red Crescent Societies.

UNDP. 2009. *Kosi Floods 2008: How We Coped! What We Need? Perception Survey on Impact and Recovery Strategies*. New Delhi: United Nations Development Programme.

UNDP and Abhiyan. 2005. *Coming Together: A Document on the Post-Earthquake Rehabilitation Efforts by Various Organisations Working in Kutch*. Bhuj: UNDP/Abhiyan.

UNDRO. 1982. *Shelter after Disaster: Guidelines for Assistance*. New York: UNDRO.

UNNATI, People in Action and CORDAID. 2012. *Participatory Assessment of Housing Vunlerability* India: UNNATI. Available at: www.peopleincentre.org/documents/hsva_training_final.pdf (accessed 21 September 2012).

Vahanvati, M. and Beza, B. 2016. Rural housing resilience in India: Is it reliant on appropriate technology or labour skills? Sustainable Futures Conference, 30 August–2 September, Nairobi, Kenya.

7

A DUTY OF CARE

Disaster recovery, community and social responsibility in Indonesia

David O'Brien, Catherine Elliott and Brendon McNiven

Introduction

On 26 December 2004. the Indian Ocean Tsunami struck coastlines in Indonesia, Thailand and Sri Lanka. Measuring between 9.1 and 9.3 on the Richter scale, the earthquake was the second largest ever recorded and centred in waters off the western coast of Aceh Province in the Indonesian island of Sumatra (see Figure 7.1). The earthquake, 150 kilometres off the coastline, shifted the seabed five metres upward in places and displaced 30 cubic kilometres of water. The resultant tsunami travelled at 1,000 kilometres per hour and rose up to 30 metres upon landfall. The cities of Banda Aceh and Meulaboh were close to being destroyed and likened to being 'hit by an atomic bomb' (ADB, 2005). With the final estimates of those killed or missing in Indonesia ranging from 170,000 to 280,000 and another 500,000 homeless, the region faced one of its darkest days.

Both the Indonesian and international communities were struck by the scenes of devastation and mobilised themselves with pledges of more than US$7 billion of aid money. More than 120 international aid agencies joined 430 Indonesian NGOs to offer immediate assistance. Within the first year of the disaster, it was estimated that approximately 120 NGOs were involved constructing new houses (Steinberg, 2007). In the years between 2005 and 2009, more than 100,000 new houses were constructed in Sumatra in direct response to the tragedy with the majority of this work being focused in the Banda Aceh and Meulaboh cities and with hundreds of other programmes along the extended coastline.

Few natural disasters have occurred at such a devastating scale, with half a million homeless and hundreds of thousands dead. Fewer still have had so many international agencies focusing so much attention in any one province. With massive resources and international teams of disaster management experts the process and outcomes should have been 'world's best practice'. Despite many accolades—much of it presented by the agencies responsible themselves—there has been an

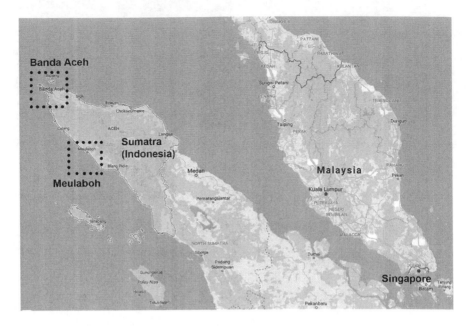

FIGURE 7.1 Map of northern Sumatra (Indonesia), locating the cities of Banda Aceh and Meulaboh, which faced the most devastation from the 2004 Indian Ocean earthquake and tsunami.

Source: Image modified from Google Maps.

undercurrent of disquiet at the lack of community engagement and consultation as well as with the final outcomes.

Planning the reconstruction process

In the tsunami's immediate aftermath and during the initial response phase, assistance was haphazard with relief agencies flying in medical aid, food, tents and clean water. Over time, the need for more coordinated efforts became pressing—particularly as the mid-year monsoon approached in 2005. In their haste, many decisions were made by the international agencies themselves with little or no consultation with the end beneficiaries of this aid. Aid agencies, competing for donors, prestige, influence and partner communities, became powerful players in a complicated sets of negotiations and began 'staking their turf' and promoting their deeds to their supporters and the media (Das, 2007; Greenblott, 2007; Steinberg, 2007).

The reconstruction effort in Indonesia was initially coordinated through the Indonesian Ministry of Public Works in partnership with the National Development Planning Agency (BAPPENAS). They stipulated that all new housing had to incorporate earthquake risk-reduction features—a wise move, given that many of the fatalities occurred from the earthquake that preceded the tsunami. However, an audit of the initial group of completed houses revealed the mantra to 'build back better' had not been followed and many houses had been built to low standards and with poor materials (Boen, 2008; Roseberry, 2008). The required infrastructure, water, electricity, sewerage and roads had not been planned for and, in many cases, the beneficiaries were unable to occupy the houses (Mudigdo, 2008; Oxfam, 2005). Partially completed houses were either abandoned or demolished, with many more requiring substantial repairs in the years that followed, and questions were raised about the lack of community engagement with a process that was largely institutionally driven (Forbes, 2006; Roseberry, 2008; Perlez, 2006).

Within four months, responsibility was shifted to a new agency dedicated to the housing recovery efforts in the affected areas. The BRR—Badan Rehabilitasi dan Rekonstruksi (Agency for Rehabilitation and Reconstruction for Aceh and Nias)—was established in April 2005, until control was handed back to the Ministry of Public Works in April 2009. This four-year period covered the vast majority of the new housing construction phase with most, but not all, of these houses built to higher standards and to improved specifications as compared to the pre-tsunami housing. The BRR first had to address a series of ethical issues and support decisions with policy and implementation teams. Should, or could, the surviving landowners be compensated for land that was lost to the sea? Was it appropriate to rebuild houses in high-risk areas? Should renters be allowed to have a house and land? Should the rich be provided with bigger houses than the poor? How big should the houses be and should there be limits on their style and amenity in order to maximise the quantity of housing?

There is an extensive body of literature that argues that these types of questions should be resolved with appropriate engagement and consultation with a range of stakeholders including the communities affected by disaster (Kenny, 2010; Lyons, 2010; Lizarralde et al., 2009; Barenstein and Leemann, 2012). However, the scale of the disaster, the loss of life, the loss of land to the sea, and the loss of land-title documents made it difficult to mobilise community decision-making efforts in the immediate aftermath of the disaster. Some less affected communities were fortunate enough to be in a position to rebuild on their own land and negotiate directly with their partner organisation to plan the reconstruction process. However, many thousands of households were displaced from their land and their extended community, and were less able to contribute to the process. The task of working with these often traumatised survivors and their immediate needs was a more complex challenge that diminished opportunities to seek community views on longer-term reconstruction needs.

Consequently the specifications developed by the BRR, on the whole, were governed by a series of principles and regulations that could never be rigidly enforced. In effect, the BRR began the reconstruction process without a mandate from all affected communities and without any comprehensive community-based aspirational vision. Local communities directly opposed many of the BRR's earlier directives, such as guidelines to maintain buffer zones between the shoreline and new housing developments. Initial decisions to provide housing only to landowners proved to be unpopular and the renters mobilised themselves until they were also offered housing—greatly increasing the number of required housing units.

FIGURE 7.2 Aerial photograph highlighting the uniformally dense nature of the Tzu Chi designed and built Great Love settlements at Panteriek (left) and Neuheun (right).

However, the BRR did lead with directives, and with few exceptions the participating aid agencies complied. Perhaps the most closely adhered to regulation was to build houses with a floor area limited to 36 square metres—generally provided within a 6×6 metre footprint. This is not a substantial house and certainly smaller than the typical urban house in Banda Aceh built before the tsunami. Both traditional and contemporary houses were designed to accommodate an extended multigenerational family. BRR guidelines generally led to the development of single-storey, masonry, bungalows, with two bedrooms, living room and outdoor kitchen/bathroom (see Figure 7.2). A comparative analysis cataloguing many of these house types was undertaken between 2007 and 2011 to showcase the range of types and the ways residents modified their houses to suit their needs (O'Brien and Ahmed, 2011, 2014).

This chapter looks at one of the largest reconstruction efforts undertaken by a single reconstruction agency taking on the responsibility to develop 3,700 houses in three locations (Panteriek, Neuheun and Meulaboh) in Indonesia. This example has been chosen not because it is the most 'typical' of the developments, but because it highlights some ethical issues not immediately apparent at other developments. This example raised the possibility that the focus on quantity, rather than quality, has contributed to some concern about the health outcomes facing residents. Can a strategy to construct larger numbers of houses, with low levels of community engagement and at lower levels of robustness and amenity, create long-term problems for recipients of reconstruction housing?

Tzu Chi and the Great Love Villages

Tzu Chi, a prominent international aid agency based in Taiwan, completed three new housing settlements in Indonesia on the island of Sumatra at Panteriek (a suburb of the provincial capital Banda Aceh), Neuheun (on the outskirts of Banda Aceh) and Meulaboh (a city 250 kilometres by road southeast of Banda Aceh). As one of the first relief agencies to respond to the earthquake and tsunami that struck the Indian Ocean on 26 December 2004, Tzu Chi played a major role in redevelopments in Indonesia, and also in Sri Lanka, maintaining its efforts with large-scale programmes over the following four years. The villages constructed by Tzu Chi in Indonesia were called 'Great Love Village I', 'Great Love Village II' and 'Great Love Village III', respectively, and were some of the largest reconstruction programmes in Aceh Province, with 700 houses constructed at Panteriek, 2,000 at Neuheun and 1,000 at Meulaboh. Another 1,000-house 'Great Love Village' was built by Tzu Chi at Hambantota in Sri Lanka. In each location, Tzu Chi also constructed schools, medical centres and community buildings. With infrastructure services like water, power and sewerage, plus roadways and drainage, these projects were significant undertakings. Many people were excited at the prospect of new housing, having lived in tents and barracks for up to two years.

Tzu Chi has achieved much renown for its programmes over many decades. From humble beginnings in 1966, a Buddhist nun, Dharma Master Cheng Yen, has led the group to its current position as one of the most significant international not-for-profit organisations. Tzu Chi has been built upon Buddhist philosophies and Cheng Yen's work with the poor in the Hualien region of Taiwan. Initially set up as a collective of housewives supporting the disadvantaged in the local neigh-bourhood, Tzu Chi's capacity to expand networks grew with the media's help and support from politicians. At the same time, an extensive grassroots supporter base of volunteers, many recruited during their time at school, helped implement and finance its charitable programmes (O'Neill, 2010).

Tzu Chi has expanded considerably over the years and now claims more than 10 million members across 47 countries, with its devotees contributing to projects that include the development and ownership of several large hospitals, schools, televisions stations, housing estates and industrial complexes. While Tzu Chi has been highly visible during international disasters and recognised for contributions after Hurricane Katrina (2005), the Indian Ocean Tsunami (2004) and the Sichuan Earthquake (2008), it has also been increasingly subjected to both domestic and international scrutiny. Allegations of using its media arm to interfere in a tainted food scandal, improper land deals, tax evasion and poor governance have been aired in the Taiwanese media (Chen, 2015; Hsu, 2015).

The houses constructed by Tzu Chi responded to the BRR's directive to not exceed the 36-square-metre footprint, but distinguished themselves from the majority of other efforts by reconstruction agencies in other ways. First, Tzu Chi constructed duplex houses—essentially joining two individual bungalows together, with the pair of houses sharing a party wall (see Figure 7.3). This was an unusual design tactic, given the existing housing typologies common in this part of Indo-nesia. However, it was an effective technique to reduce the plot of land required for each housing unit and the costs of construction, given that each house shared a wall. This decision enabled Tzu Chi to house many more families at lower costs and on smaller plots. Given the scale of Tzu Chi's Indonesian contribution, 3,700 houses, this would have been a significant factor in deliberations about the planning and delivery process. Second, the houses included a portico design that was stylistically based on a Taiwanese temple held in high esteem by the Tzu Chi Foundation—a reference made without the consent of the residents who would later occupy the houses.

It is important to note that Tzu Chi did not engage the community in the planning of the new settlements and instead constructed its houses for a 'generic' resident rather than for specific individuals, using the 'cookie-cutter' approach. Each location, at Panteriek, Neuheun and Meulaboh, had been a 'greenfield site' before Tzu Chi began construction. Although applicants for the Tzu Chi houses were interviewed several times by Tzu Chi staff about their lives before the tsu-nami, there was no discussion about the location of the settlement, its urban lay-out, the form of the houses or any symbolic imagery used to decorate the facade.

Once the construction works had been completed, Tzu Chi began to take a more active role in the lives of the people settling into their new households. As one of the largest relief agencies in the world, and with significant access to the

FIGURE 7.3 The house photographed at completion (left) is relatively modest compared to those that underwent significant additions later (right).

media and influential people, Tzu Chi managed a high-profile campaign to show-case its efforts to its Taiwanese audience. With 10 million members, Tzu Chi had a vast network of supporters and could mobilise many volunteers to participate in its relief efforts. Teams of Tzu Chi volunteers travelled to the Great Love settlements in Indonesia and Sri Lanka to coordinate community-based programmes, festivals and support groups, to offer medical support and to share their Buddhist faith.

Tzu Chi also coordinated the distribution of the housing to the beneficiaries who had been allocated to the Great Love settlements. There is no doubt that the task of allocating people to specific houses can be a complex task and requires significant resources as it is hard to please everyone. Perhaps in recognition of the task's difficulty, Tzu Chi used a randomised process, a lottery, to allocate families to houses. However, an in-depth study of the house allocation process at Tzu Chi's Great Love Village in Hambantota reveals that extended families were denied the capacity to live adjacently, despite asking Tzu Chi organisers for permission (Chen, 2014). This lack of community engagement, consultation and consideration caused deep disquiet, not least because the lottery created 'winners' and 'losers'. Those receiving a corner house on a main thoroughfare could modify their house to operate a shop or business, whereas many other houses were in more isolated loca-tions with less economic potential. Instantly, and randomly, some families' fortunes were elevated, while those of others were diminished, creating divisions within the new community.

The use of asbestos as a construction material in Great Love Villages

Another key way Tzu Chi distinguished itself from other reconstruction housing developments is more controversial, and offers further evidence that Tzu Chi operated without appropriate consultation and without the consent of the people moving to the settlements. Rather than using the ubiquitous masonry con-struction systems used by almost all reconstruction agencies, Tzu Chi constructed their houses with a lightweight construction system. External and internal wall panels were made from sheets of asbestos attached to a steel stud frame. The ceil-ing panels, plus roofing sheets, were also made from asbestos, as were the verandah linings and the curved decorative gable features to denote the entrance.

Independent analysis (tested in August 2015 by Clearsafe Environmental Solu-tions Pty Ltd, Australia) has revealed that Tzu Chi used *chrysotile asbestos*, which is one of the more common forms of asbestos used in building construction. The panels are lightweight and brittle, which makes them susceptible to damage from wear and tear, particularly in corners and areas close to doorways and windows. Any building movements that occur, such as foundation settlement, tempera-ture changes, environmental loads, such as wind, or the common earthquake movements, can fracture the panels, with microfibres breaking off into the air. Evidence of this process is easily observed on all houses in the form of cracked cor-ners and instances of 'lifting' or delamination of the asbestos sheets from the under-lying framing. The roof sheeting has performed well, whereas the asbestos wall sheeting exhibits more signs of deterioration. While serviceable at the moment, increasing levels of maintenance and repair will be required over time in order to maintain the condition of the houses.

Links between asbestos use and health

This chapter does not attempt an in-depth examination of the links between asbestos and health. Suffice to say that asbestos is known to cause mesothelioma, a form of cancer of the lungs, and prolonged inhalation of airborne asbestos fibres is known to cause serious lung-related illnesses that are ultimately fatal (Becklake, 1982). In response to the body of research that advised that building products should not contain asbestos, many western countries phased out asbestos mining and the use of asbestos products in the 1980s. Despite the widespread knowledge of the adverse effects of asbestos, its use has not been universally outlawed and it continues to be mined in a number of countries, including Russia and China, to produce building products to be used domestically and exported to countries with less stringent regulations.

Residents' awareness of asbestos use in Great Love Villages

The authors have been conducting interviews and mapping exercises at the Great Love Villages I and II during a longitudinal study stretching back to 2007. In a study of the residents' awareness of and attitudes to the use of asbestos as a construction material a series of 12 interviews (with 12 households) took place at two Tzu Chi housing complexes (Panteriek and Neuheun) during 2015. The research team included an architect/academic, an engineer and a social geographer. The focus group and individual interviews were conducted by both male and female Acehnese-speaking interpreters to maximise the capacity of interviewees to contribute their experiences.

At both locations the residents were asked 'Can you tell me about the building materials in your house?' to gauge whether residents were aware that their house was constructed with asbestos panels, and to reflect on the materials used in construction. Since moving to the Great Love houses, 11 of the 12 households have become aware that their houses are made from asbestos. None was made aware of the asbestos before moving in. The follow-up question asked residents: 'Do you have any ongoing concerns with the building materials?' The responses to this second question indicated that residents did not understand the health risks posed by the asbestos. The interviews are revealing and a summary of some of the conversations is outlined below to provide some insights into residents' attitudes to the use of asbestos in their housing.

- Nurhayati and her husband live in a house with their four school-age children. Nurhayati knows that the house has asbestos: '[There is] no serious problem with asbestos . . . the house is strong, [I have] lived here for many years and not had any problems.' She explained that there is nothing she doesn't like about the house; although she pointed out that the ceiling is a little broken, she said that her husband could fix it. Nurhayati explained that for herself and her husband, the important thing is to have a house, because many others do not.

- Sri Iza said that her house was made from gypsum: 'However, this house causes asthma'. She claimed that one advantage of this material is that it is not easy to burn. She had a bad experience when a timber house burned down.
- Denny explained that he would have preferred a house made from another type of material that is thicker and less dusty than asbestos. He described how asbestos and gypsum are almost the same in terms of their positives and negatives but that they both have the 'dust effect'.
- Abdullah Umar turned down an opportunity to live in a masonry Habitat house because he was impatient, but he now regrets that decision because he has become aware that the Tzu Chi houses are made from asbestos. Had he known, he would have waited for the Habitat house. 'We are grateful but this is not aid [housing], it only lasts for 20–30 years. [But] in my opinion the houses which are the worst are the Tzu Chi houses, because of the asbestos; the standard for other houses is brick. This is the worst NGO . . . We are grateful [to have a house] but not satisfied'. Abdullah did not mention health concerns.
- Alek and Siti live with their one son and know that the house is made from asbestos. Siti said that she had heard that asbestos is a temporary material (not durable) and that it is not good for their health to use it for a long time. She also said that it was easy to crack and pointed to the fissures between the walls and ceiling panels and where parts of the ceiling had fallen down.
- M. Zein asked the researchers whether asbestos affects health, because he had heard it creates breathing problems.

Information taken from the interviews shows that while many of the occupants of the houses were aware that they were constructed from asbestos, only two mentioned that it may have health effects. There was no knowledge of the mechanisms behind the dangers (airborne particles) and consequently no understanding of measures that may be taken to avoid them.

Activities that inflame risk

It must be accepted that the residents of these Great Love Villages are at a significantly heightened risk of contracting mesothelioma and other lung diseases due to their increased exposure to asbestos fibres in their housing materials. Unfortunately, the risk is exacerbated by two key activities that are commonly occurring in the houses. First, as the interviews attest, the houses are dusty, both internally and externally, as the fibres spall off the asbestos panels and become airborne. The problem of spalling is gradually increasing as the painted surface of the panels deteriorates with further exposure of the asbestos fibres to the general atmosphere. The white paint, initially applied at the time of construction by the Tzu Chi contractors, was not high quality and needs additional coats, and a strict maintenance schedule, to keep a workable seal to the surface of the asbestos. Unfortunately, it is difficult to obtain quality paint in Indonesia due to cost and quality-control issues, hence the painted surfaces at the Great Love Villages do not retain their robust 'skins' or clean appearance.

The second issue that exacerbates the risk factors is associated with the modification and rebuilding programmes that have taken place in the Great Love Villages in the years since completion. Surveys of the housing has revealed that the residents accommodated in the majority of the houses built by reconstruction agencies after the Indian Ocean Tsunami have taken responsibility for modifying their houses, typically by building additional rooms, adding decorative elements, or demolishing the houses completely and building new ones from scratch (O'Brien and Ahmed, 2011). Since the footprint is only 36 square metres (much smaller than the typical pre-tsunami housing), many households, particularly in the Panteriek community, have made efforts to significantly improve their housing status. Some have demolished the Tzu Chi houses and built entirely new masonry dwellings. In a site survey conducted in 2015 at the Panteriek community, an estimated 700 houses, approximately 50 per cent of the total, have undergone significant modifications that have included the removal of some of the asbestos walls, while around 8 per cent of the houses have been demolished and replaced.

The process of demolishing a full Tzu Chi house and rebuilding with masonry is an aspirational, but achievable, step in a community such as Panteriek, which has a reasonable number of relatively affluent households as compared to many other communities in Indonesia. One can only anticipate that this trend will increase in the future as Indonesia's growing middle class and upwardly mobile households understandably wish to improve their living conditions with housing that reflects their aspirations and improving social status. However, this process requires demolition of the original Tzu Chi house, with its asbestos, and this is where the risks to the community's health are heightened. There is no evidence that the demolition process is managed to reduce the spread of airborne fibres, resulting in large amounts of asbestos dust being distributed around and beyond the construction zone. Adding to the problems facing many construction sites are the piles of discarded asbestos left by the roadside with no clear strategy to reduce the associated dust or dispose of the panels in a safe way.

A duty of care

There is no written documentation to explain why Tzu Chi took the decision to use asbestos despite the international community making significant steps to outlaw its use in construction. Perhaps Tzu Chi wished to reduce construction costs and the time required to produce so many houses. However, ignoring the issues raised in this chapter is, we argue, morally unacceptable, given the ongoing health risks to the communities residing in these Great Love Villages. Tzu Chi should take the lead to remedy these immediate risks to the villages, re-engage with the communities, and review their decision not to disclose the use of asbestos at the beginning of their intervention in Indonesia.

It would be naïve to believe it is possible to remove all of the risks posed by asbestos to the residents of the Great Love Villages as well as people from neighbouring villages, casual visitors to the communities, construction workers and all those people who come into contact with the asbestos waste once it has been removed from the sites and dumped. An inconceivable number of people have already been, and will continue to be, exposed to these fibres. However, it is unacceptable to ignore the issue and leave the residents of these communities unaware of the risks they face in their everyday lives.

Given the low level of understanding about the risks associated with asbestos within the three affected communities the question must be posed: how should the residents of the Great Love Villages be informed of the health risks in order to change their behaviour without inflicting undue stress and anxiety? Given the scale of the problem we must also ask: who should take responsibility for informing the residents and who should take responsibility for remedial works? It is beyond the scope of this chapter to answer these questions but they do lead to another that shall be addressed in the remainder of this section. Given the issue at hand and the fact that no other agencies are recognising the problem or addressing the key issues, what, in the more immediate term, can be done to reduce the risks for the residents of the Great Love Villages? Some measures can and should be put in place and the responsibility for doing this should rest with Tzu Chi. In the immediate term these measures should, at the bare minimum, include the following:

- Free access to paint or sealing agents, plus appropriate application equipment, to contain the dust spalling from the asbestos panels.
- The safe removal of asbestos from public buildings constructed by Tzu Chi and replacement with non-asbestos panels.
- Free education programmes for construction workers informing them of best-practice methods for asbestos removal, plus free access to protective clothing and breathing apparatus when working with asbestos.
- The identification of suitable waste storage plants for discarded asbestos panels.
- Immediate review of any policies that encourage the use of asbestos.
- Immediate review of Tzu Chi's duty of care policies.

Tzu Chi, an organisation that prides itself on its community engagement, has ultimately let down the residents of the Great Love communities—first, by not involving them in the design of the Great Love Villages and, second, by not making it clear that the houses are constructed with asbestos. It is now time for Tzu Chi to re-engage with the residents of the Great Love Villages, accept responsibility, and determine an appropriate path to address the health risks facing the communities in the future.

References

ADB (2005) *From Disaster to Reconstruction: A Report on ADB's Response to the Asian Tsunami*. ADB (Asian Development Bank), Manila.

Barenstein, Jennifer and Leemann, Esther (2012) *Post Disaster Reconstruction and Change: Communities' Perspectives*. CRC Press, Florida.

Becklake, Margaret (1982) 'Asbestos-related Diseases of the Lungs and Pleura', *American Review of Respiratory Disease*, 126(2), pp. 187–194.

Boen, Teddy (2008) 'Reconstruction of Houses in Aceh, Three Years after the December 26, 2004 Tsunami', *Proceedings of the International Conference on Earthquake Engineering and Disaster Mitigation, Jakarta.*

Chen, Christie (2015) 'Tzu Chi Row Reflects Three Social Changes', *Taiwan News.* Available at: www.taiwannews.com.tw/etn/news_content.php?id=2696775 (accessed 1 August 2016).

Chen, Ted (2014) 'Religious NGO Approaches to Post-Disaster Housing Reconstruction in Sri Lanka', unpublished PhD, University of Melbourne.

Das, Joyati (2007) 'Reconstructing Responses: The Trials and Tribulations of World Vision's Tsunami Shelter Program', *Annual Review 2007*, World Vision, Australia.

Forbes, Mark (2006) 'Aid Agencies "Lied" over Tsunami Aid', *The Age*, 9 March.

Greenblott, Kara (2007) *Shelter Programming: Learning from the Asia Tsunami Response.* World Vision, London.

Hsu, Elizabeth (2015) 'Why Tzu Chi is Sparking Resentment', *Taiwan News.* Available at: www.taiwannews.com.tw/etn/news_content.php?id=2698712 (accessed 1 August 2016).

Kenny, Sue (2010) 'Reconstruction through Participatory Practice?' in M. Clark, I. Fanany and S. Kenny (eds) *Post-Disaster Reconstruction: Lessons from Aceh.* Earthscan, London.

Lizzeralde, G., Cassidy, J. and Davidson, C. (2009) *Rebuilding after Disasters.* Routledge, London.

Lyons, Michal (2010) 'Can Large-Scale Participation Be People Centred? Evaluating Reconstruction as Development', in M. Lyons, T. Schilderman and C. Boano (eds) *Building Back Better: Delivering People-centred Housing Reconstruction at Scale.* Practical Action Publishing, Warwickshire.

Mudigdo, Adi (2008) 'Review of Tsunami Housing in Aceh' (personal report), Jakarta.

O'Brien, David and Ahmed, Iftekhar (2011) 'Donor-Driven Housing, Owner-Driven Needs: Post-Tsunami Reconstruction in Aceh, Indonesia', in N. Kaufman (ed.) *Pressures and Distortions: City Dwellers as Builders and Critics.* Rafael Vinoly Architects, New York.

O'Brien, David and Ahmed, Iftekhar (2014) 'Global and Regional Paradigms of Reconstruction Housing in Banda Aceh', *Open House International*, 39(3), pp. 37–46.

O'Neill, Mark (2010) *Tzu Chi: Serving with Compassion.* John Wiley & Sons, Singapore.

Oxfam (2005) *A Place to Stay, a Place to Live: Challenges in Providing Shelter in India, Indonesia and Sri Lanka after the Tsunami.* Oxfam, Oxford.

Perlez, Jane (2006) 'Aid Groups Are Criticized over Tsunami Reconstruction', *New York Times*, 27 July. Available at: www.nytimes.com/2006/07/27/world/asia/27indo.html (accessed 19 April 2017).

Roseberry, Rachel (2008) 'A Balancing Act: An Assessment of the Environmental Sustainability of Permanent Housing Constructed by International Community in Post-Disaster Aceh'. *Proceedings of the 4th International i-Rec Conference, University of Canterbury, Christchurch, New Zealand.*

Steinberg, Florian (2007) 'Housing Reconstruction and Rehabilitation in Aceh and Nias, Indonesia: Rebuilding Lives', *Habitat International*, 31(1), pp. 150–166.

8

THE ROLE OF COMMUNITY ENGAGEMENT IN POST-DISASTER RECONSTRUCTION

The cases of L'Aquila and Emilia Romagna, Italy

Lorenza Lazzati

Background

The inclusion of participation in government policies and plans is relatively widespread. However, the premise for a truly shared decision-making power between authorities and community resides in the institutionalisation of participation (Ackerman, 2003). In the case of Italy, participation has been progressively introduced in local and regional administration in different areas, such as budgeting and planning (Sbilanciamoci!, 2010). Still, such initiatives are not consistent throughout the country. Italian governance is based on the decentralisation of power, and matters like planning and healthcare are regionally regulated and administered.

Regarding post-disaster response strategies since the 1950s, Italy has adopted case-by-case solutions, wherein participation has had different roles. The form of recovery has shifted between top-down state-centred approach and a more inclusive and participatory approach (Clementi and Fusero, 2011). On each occasion, the legislator has granted ad hoc emergency powers to the concerned authorities, be they local, regional or national, allowing them to bypass normal legislation (Özerdem and Rufini, 2012). Thus the role of participation has changed, based on the organisational structure of the authorities overseeing the reconstruction. One principle which has consistently underpinned the decisions made so far at national level is that of rebuilding 'where it was, the way it was' (Nerozzi and Romani, 2014).

To date, the regional responses have proven to be more successful, as in the reconstructions of Friuli (1976) and Umbria (1997), while state-led reconstructions have been prone to politicisation. Such was the case of Irpinia (Campania) in 1980, which resulted in fund misallocation and one-third of the affected population living in pre-fabricated units or containers for over a decade (Özerdem and Rufini, 2012).

This chapter reports on research which investigated the extent to which a people-centred reconstruction approach can contribute to the development of resilience of communities affected by earthquakes in Italy. The research considered the intermediate recovery phase, specifically the process around the design and creation of physical (urban) planning documents that would integrate with the existing physical planning and would inform the reconstruction decisions. For this purpose, two case studies have been selected, L'Aquila and Emilia Romagna.

The definition of resilience used in this research has been derived from Folke *et al* (2005) ecological studies on the *social-ecological system* (SES). The SES definition of resilience, as applied to post-disaster situations, has three main characteristics: firstly, the level of impact the community can absorb while maintaining its social cohesion and identity; secondly, the extent to which the community can become self-reliant; thirdly, the degree to which the community is able to learn and adapt to greater future impacts (Folke *et al.*, 2005).

The role of the Department of Civil Protection (DPC)

Disaster response in Italy has always relied on the Civil Protection Service. However, the department's organisational structure, responsibilities and competencies were only established in 1970 (DPC, 2014a). Subsequently, the 1980 earthquake in Irpinia highlighted the need for radical improvement in efficiencies and coordination within the system.

A significant reform began in 1992 and continued into the mid-1990s, when the DPC was reorganised from a central structure, coordinated by the Ministry for Internal Affairs, to a decentralised one, where regional and local authorities had greater responsibility and authority in prediction, prevention and mitigation of possible disasters (DPC, 2014a). However, in 2001 National Law 401/2001 inverted this process and bestowed all competencies to the President of the Cabinet of Ministries, who could exercise his/her power through the DPC (Legge n. 401/2001).

To strengthen the DPC's authority even further, the national government brought 'major events' (*grandi eventi*) under the competency of the department. This centralisation of power and responsibilities was a significant change in the role of the DPC. Under this new legal framework, the national government determined what qualified as a 'major event' and the DPC could bypass existing planning, environmental and heritage legislation to organise and build the structure necessary for said event. In 2009, the government extended the notion of building for major events to include even private works, such as housing construction, where as before it had been limited to providing public works (OPCM, 2009).

By 2010, the DPC had almost unlimited power and had to answer only to the President of the Cabinet of Ministries. As discussed later in this chapter, this legal framework allowed the reconstruction of L'Aquila to develop as it did in 2009. With the change of government at the end of 2011, new legislation was introduced to revert the DPC to a structure and competency similar to that of the 1990s. Major works were no longer under the responsibility of the DPC, and National Law 100/2012 introduced a 90-day time limit to the emergency. The reconstruction in Emilia Romagna took place under this new legal framework (Legge n. 100/2012).

The study

The study involved the analysis of administrative documents, legislation, formal studies, media releases and filmed documentaries. The limited timeframe for the research posed a constraint on the resources that could be easily and rapidly acquired. Whenever possible, official governmental and legislative documents were consulted to avoid possible bias.

The focus of the study was to ascertain the level of community participation in the creation of planning documents specifically designed for the reconstruction and whether such engagement is capable of creating an informed design that contributes to building community resilience. To this end, the research analysed the

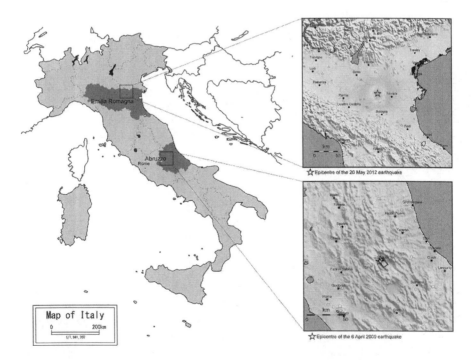

FIGURE 8.1 Map of Italy: earthquake epicentres.

Source: Shiro chizu senmon ten. Available at: www.english.freemap.jp/item/europe/italy.html (accessed 14 February 2016). Epicentre maps source: U.S. Geological Survey, Department of the Interior.

type of engagement (i.e. participation versus consultation) carried out by institutions/concerned authorities when creating reconstruction plans.

L'Aquila and Emilia Romagna were chosen as case studies for two reasons. First, they are close in time: the L'Aquila earthquake was in 2009 and the one in Emilia Romagna occurred in 2012. Second, the two cases differ in their urban structure, geographical, political and historic conditions. Furthermore, the reconstruction responses adopted in the two cases are exact opposites.

L'Aquila case study

Context

L'Aquila is the regional capital of Abruzzo, a region located in the geographical centre of Italy. Despite its location, Abruzzo is considered part of the Mezzogiorno (the south) from a cultural and socio-political perspective. The QUARAS report (Sbilanciamoci!, 2010) considers Abruzzo as the region with the best quality of development in the Mezzogiorno, and places it 12th out of 20 regions. The region performs best in the area of environment, but poorly in the area of participation, with scarcely any citizen involvement in civic organisations (8.1 per cent) or voluntary activities (2.2 organisations for every 10,000 inhabitants) (Sbilanciamoci!, 2010).

Prior to the 2009 earthquake, the urban structure of the area of L'Aquila was characterised by a monocentric layout that reflected the relationship between the city suburbs and peripheral towns and L'Aquila's historic city centre (Comune di L'Aquila, 2011). L'Aquila hosted multiple functions, such as regional and local government administrations, universities, cultural institutions, professional and commercial services, as well as residential housing. The presence of these sectors gave the feeling of an active and lively city whose value was increased by the quality of its historical architecture. Up to the time of the earthquake, the city was the focal point of the community's social and cultural life. It had experienced demographic growth, while the populations of the peripheral towns had stagnated (Comune di L'Aquila, 2011).

The immediate response in L'Aquila

The 2009 earthquake occurred on 6 April 2009. Fifty-seven towns and 65,000 people were affected, with a death toll of 308 people. The earthquake extensively damaged the city and the surrounding towns. It was reported that 33 per cent of the existing buildings were declared completely unusable (Banca d'Italia, 2013).

The immediate relief was carried out by the DPC, which also coordinated the action of the fire brigade, the army and other organisations that provided shelter and relief. A month after the earthquake, 32,000 people were accommodated in 170 tent camps built by the DPC, while the remaining 33,000 people were accommodated in hotels along the coast (approximately 200 kilometres from L'Aquila) and in private residences (DPC, 2009).

Immediately after the earthquake, the national government declared a state of emergency to last nine months, until 31 January 2010. During this period, the head of the DPC acted as the delegated commissioner for the response. At the same time, L'Aquila's mayor declared the entire historic town centre and 49 other historic centres within the municipality as 'red zones' (Città dell'Aquila, 2014), which were restricted to authorised personnel only, with army checkpoints set up at every point of entry (*Draquila*, 2010).

The reconstruction process: MAP and CASE projects

In its early phases, the reconstruction was heavily politicised by the national government, while the Mayor of L'Aquila, the citizens and their representatives were not consulted and were excluded from any decision making (*Draquila*, 2010). The reconstruction of the existing city centre and other smaller towns was not considered, despite their social, cultural and historical value. The top-down approach adopted in L'Aquila was reminiscent of old-style welfarism, with the community a passive recipient (Puglielli, 2010).

The national government promoted the provision of new temporary and transitional accommodation (new towns) to respond to the immediate housing need generated by the extensive damage to the existing buildings. The government's effort focused only on housing supply and lacked a strategic plan for the reconstruction of the city in its entirety (Fusero, 2011). To gain public support for the new towns, the government promoted the message that they would provide houses and not containers, referring to unsuccessful previous solutions adopted in cases like Irpinia (Campania) (*Draquila*, 2010). The decision to build temporary and transitional accommodation was announced by the national government two weeks after the earthquake, and at the end of September, just five months later, the DPC started to hand over the first houses (DPC, 2014b).

The legislative framework allowed the DPC to act as a developer for the construction of the new towns, while overruling regional and local governments (Özerdem and Rufini, 2012). The DPC chose the areas for the construction of the residential units in close proximity to existing *frazioni*[1] with the exception of Sant'Elia and Sant'Antonio. The sites were mostly zoned either as agricultural areas or as environmentally protected areas (Clementi and Fusero, 2011).

The intervention implemented by the DPC consists of MAP housing units (*Moduli Abitativi Provvisori*) and CASE (*Complessi Antisismici Sostenibili Ecocompatibili*). The MAP project consists of 107 individual timber houses built in the smaller towns around L'Aquila (Fusero, 2011). The CASE intervention, on the other hand, consists of 185 pre-fabricated three-storey buildings, a total of 4,600 apartments, constructed across 19 areas around L'Aquila and its suburbs. These apartments were built to a high specification in terms of anti-seismic technology and are environmentally sustainable (DPC, 2014b). They were fully furnished and supplied with appliances and basic items such as bedding and kitchenware.

However, these housing solutions can be considered temporary accommodation only in the sense that their occupants are temporary residents. The buildings themselves are very much permanent, since the type of construction that was used guarantees a long life beyond the emergency period (Fusero, 2011). Moreover, these developments lacked social infrastructure, such as schools and childcare facilities, as well as adequate public transport. There were plans to build shopping malls near the new houses, but this urban model of suburban houses built around a shopping mall is foreign to most of these areas (*Draquila*, 2010).

Prior to the earthquake, the city and its suburbs were laid out in a monocentric urban structure, where the historic centre played a pivotal role as the heart of the community and favoured the integration of commercial and residential use. The construction of the new towns changed that structure to a polycentric one and disregarded the need to repair the historic centre (Comune di L'Aquila, 2011). To this day, large parts of the historic city centre are still not accessible.

The post-DPC period

In 2011, the DPC handed over the coordination of the reconstruction process to the regional and local authorities. The local governments inherited a city with unresolved earthquake-related issues, large areas were still considered red zones, and the new towns, which were effectively dormitory towns, added new urbanisation issues (Fusero, 2011; Comune di L'Aquila, 2011).

In contrast to reconstruction implemented by the national government and the DPC, the 'Plan for the Historic Town Centres of L'Aquila and Suburbs' (*Piano dei centri storici de L'Aquila e frazioni*), adopted by the Municipality of L'Aquila in 2011, recognises the crucial social and cultural importance of the historic city centre in the community's identity. This change in direction is indicative of the difference in view between the local authorities and the national government. While the national government had addressed the immediate housing need, it had disregarded intervention pertaining to the existing city centre. A more inclusive process would have highlighted the central role of the historic centre of L'Aquila. In addition, in 2012, the Municipality of L'Aquila adopted the 'Regulation for the Participation of Citizens' (*Regolamento sugli Istituti della Partecipazione*), with the intention of developing a more active and cohesive society (Città dell'Aquila, 2012).

Emilia Romagna (Finale Emilia) case study

Context

Emilia Romagna is a region in the north of Italy. Over the past 30 years, the region's economic profile has changed from agricultural to industrial, mostly machinery manufacturing (Laghi *et al.*, 2013). The area affected by the 2012 earthquake, commonly referred to as '*la Bassa*', is where most of the industries are located: it generates 2 per cent of the national GDP (Regione Emilia Romagna, 2014).

La Bassa has been historically characterised by polycentric urbanisation established over many centuries. The historic town centres, once the focal points of the local communities' cultural identity, were progressively abandoned in the face of industrialisation and urban sprawl. This was considered to be one of the causes of their socio-economic vulnerability, which was exacerbated by the earthquake (Nerozzi and Romani, 2014; Tortoioli, 2014; Zanelli, 2014). Nonetheless, the QUARAS report (Sbilanciamoci!, 2010) classifies Emilia Romagna as the second-best region for quality of development. The region performs particularly well thanks to its efficient social service system and healthcare. In addition, it has an active and engaged civil society, deeply rooted in its history, and a large number of voluntary associations.

Participatory approach in the planning system of Emilia Romagna

The planning system is a reflection of the civil society of the region. Since the 1950s, the neighbourhood units have been actively working for the renovation of Bologna, the regional capital. However, the institutionalisation of participation in planning did not happen until the 2000s through Regional Laws n. 6/2009[2] and n. 3/2010.[3] The former aims to bring participation into planning decision making, as well as implementation. The latter changed the role of citizens' participation from a consulting role to one of active co-producer of regional policies (Proli, 2011).

To foster participation, the regional government published guidelines to support local authorities and citizens' organisations who want to use a participatory approach to engage in community life, not just planning. It is worth noting that the guidelines stress the importance of an interdisciplinary approach, which can, in the view of the legislator, foster more effective decisions (Regione Emilia Romagna, 2009).

The immediate response in Emilia Romagna

The earthquake in Emilia Romagna registered two main shocks. The first occurred on 20 May 2012 and the second was on 29 May, affecting a total of 58 municipalities, displacing 45,000 people across five provinces. The human toll was 29 people killed with 390 injured (Regione Emilia Romagna, 2014). The national government declared a state of emergency to last 12 months, until 31 May 2013 (Decreto Legge n. 74).

The change in the legislative framework of the DPC, which brought its function back to that of the late 1990s, dramatically affected the role and involvement of the department in the reconstruction. In Emilia Romagna, the DPC was to provide immediate relief, but it was not responsible for the construction of houses, as it had been in L'Aquila. From the very beginning, the Emilia Romagna response took an opposite stance to the L'Aquila response. The chronological closeness of the two events meant that L'Aquila's shortcomings, where the city centre was still inaccessible and the new towns had modified the urban structure, were still fresh in the public's mind (ANCSA, 2013).

The reconstruction process

The national government decreed that the role of commissioner for the reconstruction was to be given to the President of the Region (Decreto Legge n. 74), thus conferring decisional power regarding the reconstruction process on to him. The regional coordination permitted a localised response that focused on the specificity of the affected areas and communities.

The regional government approved new legislation to regulate and manage the response (Legge Regionale n. 16/2012). This regional law introduces the Reconstruction Plan (*Piano della Ricostruzione*, or PdR[4]), a planning document that integrates with the existing municipal strategical urban planning. The objective of this law is to repair the existing heritage sites while promoting a sustainable and energy-efficient reconstruction, favouring the creation of services and open spaces that reflect the cultural identity of each town. In the rural areas, the objective of the law is to promote the recovery of agricultural use and related activities as well as the reconstruction of the existing buildings, even if they are no longer connected to agricultural activities. Legge Regionale n. 16/2012 recognises the cultural and environmental value of the rural landscape and the agricultural heritage.

Furthermore, the law recognises the importance of stakeholders' participation in enhancing the effectiveness of the PdR (Legge Regionale n. 16/2012). The type of community engagement suggested by the law requires a collaborative approach between administration and community to achieve a common goal. Community engagement is developed through the use of public assemblies and/or public forums to define the intervention's objectives and workshops to create physical planning design solutions. Focus groups, on the other hand, assess groups' positions on different topics while site visits are used to enhance people's knowledge of their own territory (Regione Emilia Romagna, 2009). The law also makes provision for participatory monitoring in order to ensure transparency, accountability and the continuous involvement of all stakeholders (Legge Regionale n. 16/2012).

The town of Finale Emilia illustrates how the process was applied. It was heavily damaged by the earthquake (ANCSA, 2013), and at the time of the research the municipality had just completed the engagement process. Here, the municipality articulated the community engagement in two phases: planning and design. During the planning phase, which took place between September and December 2013, the citizens were asked to identify critical issues related to the urban centres. The themes that emerged during the planning phase were incorporated into the PdR. Therein, they are referred to as areas that require specific planning (*ambiti di approfondimento progettuale*). The objective of the community engagement during the design phase was to document the prioritised requirements that will inform the design of the specific projects in the planning phase of the engagement (Comune di Finale Emilia, 2014).

FIGURE 8.2 Torre dei Modenesi, Finale Emilia.

Source: Paolo Bona/Shutterstock.com.

Through the engagement process, the community identified as critical issues the need to reorganise existing infrastructure, particularly the bus station and the sports centre, and to promote the requalification of the *frazioni* and the industrial areas. As a design proposal, the community agreed to renovate the town centre in order to develop socio-economic and cultural activities. Priority was given to the reconstruction of historic buildings and monuments, since they represent a familiar landscape for the community (Comune di Finale Emilia, 2014).

At the time of research, it was too early to establish how the participatory process had influenced the reconstruction of the affected areas. Nonetheless, the creation of an ad hoc legislative framework ensures that each municipality is able to engage with its own community and address specific issues and priorities. In addition, the regional government had allowed for the PdR to be phased in order to meet the more urgent needs and critical issues (Nerozzi and Romani, 2014). At the time of research, 25 of the 58 affected municipalities had adopted a PdR and the participatory process was at various stages in different towns.

Conclusion

This chapter presented two post-disaster case studies in Italy, L'Aquila and Emilia Romagna. In the case of L'Aquila, the reconstruction process did not follow the principle of rebuilding 'where it was, the way it was' (Nerozzi and Romani, 2014) that informed previous disaster responses in Italy. From the very beginning, the heavy politicisation of the process steered the national government's decisions to support the construction of the new towns, excluding the local communities from the decision-making process. The result of this early decision fragmented and dispersed the local population across several suburbs, new towns and towns along the coast. Furthermore, the construction of the new towns led to a polycentric urban development process that was foreign to the area and to the community's way of living, which had previously gravitated around L'Aquila city centre. When the regional and local governments took over responsibility for the reconstruction from the national authorities, they inherited a territory which not only had unresolved disaster-related issues but was also aggravated by new urbanisation issues related to the poorly planned new towns.

The reconstruction in Emilia Romagna, on the other hand, was based on a new principle that combines the reconstruction and renovation of affected areas (Cocchi, 2014). The legal framework for the response sits within Emilia Romagna's regional planning system, which adopted the participatory process in 2009. Undoubtedly, the familiarity of the Emilia Romagna regional and local governments with community engagement practices in advance of the earthquake was an advantage that L'Aquila's local government did not have. The fore knowledge of participatory practices meant that the authorities did not have to learn and apply new practices under the pressure of a post-disaster response. Although the ultimate decision-making power still remains with local institutions, the people-centred approach adopted in Emilia Romagna provided the community with the opportunity to be heard.

The contrasting findings of the two case studies confirmed that an active community engagement can contribute to making communities more resilient. In Emilia Romagna, the reconstruction guideline documents that resulted from the early engagement have the potential to express the communities' cultural identities. These documents have been created through the collaborative efforts of the community and local authorities, as in the case of Finale Emilia. Here people's involvement has helped to identify priorities and needs, formulating a vision for the long-term development of their town, thus fostering social cohesion and a sense of identity.

Conversely, the centralised top-down approach adopted in L'Aquila turned the community into a passive recipient, undermining people's self-reliance and resilience. While it addressed the immediate housing need, the reconstruction led by the national government and the DPC in L'Aquila did not foster community cohesion and disregarded existing social networks and interactions, as well as community characteristics and capacity. Furthermore, the possibility of the community being involved in the decision-making process can help mitigate the risk of politicisation and exclusionary decision making, enhancing government accountability.

At the time of research, the Emilia Romagna reconstruction process was still in its early stages. Although the legal framework and early engagement process have the potential to develop community resilience, further research should be conducted to assess the long-term impact of the reconstruction response.

Notes

1 A *frazione* is a small town, akin in size to a parish or village in the UK, with administrative functions located in a larger adjacent town.
2 *Governo e riqualificazione solidale del territorio* [Governance and supportive regeneration of the territory], 6 July 2009.
3 *Norme per la definizione, riordino e promozione delle procedure di consultazione e partecipazione alla elaborazione delle politiche regionali e locali* [Norms for the definition, rearrangement and promotion of consultation and participation procedures to the elaboration of local and regional policies], 9 February 2010.
4 The *Piano della Ricostruzione* (PdR) is a planning document introduced by Legge Regionale n.16/2012 to support the municipalities affected by the 2012 earthquake. The PdR, by integrating existing programmatic urban planning, allows an individual municipality to identify, design and intervene in areas/buildings damaged by the earthquake based on the local situation and needs. The design of such planning documents is subject to prior community consultation.

References

Ackerman, J. (2003) 'Co-Governance for Accountability: Beyond "Exit" and "Voice"', *World Development*, 32(3), pp. 447–463.
ANCSA (2013) *Le forme della ricostruzione – Terremoto Emilia Romagna*. Città di Castello (PG): Associazione Nazionale Centri Storico-Artistici and Regione Emilia Romagna.
Banca d'Italia (2013) *Economie Regionali, l'economia dell'Abruzzo*. Banca d'Italia – Eurosistema. Available at: www.bancaditalia.it/pubblicazioni/econo/ecore/2013/analisi_s-r/1412_abruzzo/1413_abruzzo.pdf (accessed 10 September 2014).
Città dell'Aquila (2012) *Regolamento sugli Istituti di Partecipazione*. Città dell'Aquila, allegato alla Delibera Consiglio Comunale, n. 13, 26 January 2012.
Città dell'Aquila (2014) *Settore ricostruzione private, settore ricostruzione pubblica* (slide presentation). Città dell'Aquila: Assessorato alla Ricostruzione. Available at: www.comune.laquila.gov.it/pagina818_cinque-anni-dopo-2009–2014.html (accessed 16 September 2014).
Clementi, A. and Fusero, P. (eds) (2011) *Progettare dopo il terremoto: esperienze per l'Abruzzo* [Designing after the earthquake: the Abruzzo region experience]. Barcelona: Actar.
Cocchi, E. (2014) 'Editoriale', *Inforum*, 45, p. 2. Regione Emilia Romagna. Available at: http://territorio.regione.emilia-romagna.it/entra-in-regione/riviste-e-pubblicazioni/inforum/anno-2014/INFORUM_45_maggio_2014.pdf (accessed 12 September 2014).

Comune di L'Aquila (2011) *Il piano di ricostruzione dei centri storici di L'Aquila e frazioni. Linee di indirizzo strategico.* L'Aquila: Assessorato alla Ricostruzione e Pianificazione, Settore Pianificazione e Ripianificazione del territorio. Available at: www.comune.laquila.gov.it/pagina200_le-linee-di-indirizzo-strategico.html (accessed 16 September 2014).

Comune di Finale Emilia (2014) *Documento di proposta partecipata.* Comune di Finale Emilia. Available at: http://nuovosito.comunefinale.net/images/DocPP_Finale_Emilia.pdf (accessed 12 September 2014).

Decreto Legge n. 74, *Interventi urgenti per le popolazioni colpite dagli eventi sismici nelle province di Bologna, Modena, Ferrara, Mantova, Reggio Emilia e Rovigo il 20 e il 29 Maggio 2012,* 6 June 2012.

DPC (2009) *Alcuni dati sulla popolazione assistita e sulle verifiche di agibilità.* Dipartimento della Protezione Civile. Available at: www.protezionecivile.gov.it/jcms/it/view_new.wp?facetNode_1=data new%282009%29&prevPage=news&contentId=NEW496 (accessed 7 September 2014).

DPC (2014a) *La protezione civile nella storia.* Dipartimento della Protezione Civile. Available at: www.protezionecivile.gov.it/jcms/it/storia.wp (accessed 7 September 2014).

DPC (2014b) *Dossier – CASE – Complessi Antisismici Sostenibili ed Ecocompatibili.* Dipartimento della Protezione Civile. Available at: www.protezionecivile.gov.it/jcms/it/view_dossier.wp?contentId=DOS274 (accessed 7 September 2014).

Draquila: L'Italia che trema (2010) (documentary). Rome: Secol Superbo e Sciocco Produzioni, Gruppo Ambra, Alba Produzioni.

Folke, C., Hahn, T., Olsson, P. and Norberg, J. (2005) 'Adaptive Governance of Social-Ecological Systems', *Annual Review of Environment and Resources*, 30(1), pp. 441–473.

Fusero, P. (2011) 'Piano CASE e MAP', in A. Clementi and P. Fuser (eds) *Progettare dopo il terremoto: esperienze per l'Abruzzo* [Designing after the earthquake: the Abruzzo region experience]. Barcelona: Actar, pp. 190–199.

Laghi, A., Mancini, M., Piergiovanni, R., Parisi, V., Briolini, M., De Siena, D. and Migliardo, S. (2013) *La struttura imprenditoriale e produttiva dell'Emilia-Romagna.* Regione Emilia Romagna: Archivio Statistico delle Imprese Attive (ASIA), ISTAT, SISTAN. Available at: http://statistica.regione.emilia-romagna.it/allegati/pubbl/ASIA2010/view (accessed 23 September 2014).

Legge n. 100/2012, *Conversione in legge, con modificazioni, del decreto legge 15 maggio 2012, n. 59, recante disposizioni urgenti per il riordino della protezione civile* [Conversion into law, with modifications, of the Decree Law 15 May 2012, n. 59, on urgent measures for the reorganization of the civil protection], 12 July 2012.

Legge n. 401/2001, *Legge di conversione del decreto legge 7 Settembre 2001, n. 343, Disposizioni urgenti per assicurare il coordinamento operativo delle strutture preposte alle attività di protezione civile e per migliorare le strutture logistiche nel settore della difesa civile* [Law to convert the Decree Law 7 September 2001, n. 343, Urgent disposition to ensure the operational coordination of the departments involved in civil protection activities], 9 November 2001.

Legge Regionale n. 16/2012, *Norme per la ricostruzione nei territori interessati dal sisma del 20 e 29 maggio 2012,* 21 December 2012.

Nerozzi, B. and Romani, M. (2014) 'Il Piano della ricostruzione: un nuovo approccio disciplinare e metodologico', *Inforum*, 45, pp. 12–15.

OPCM (2009) *Ordinanza della Presidenza del Consiglio dei Ministri del 30 Giugno 2009,* n. 3787.

Özerdem, A. and Rufini, G. (2012) 'L'Aquila's reconstruction challenges: has Italy learned from its previous earthquake disasters?', *Disasters*, 37(1), pp. 119–143.

Proli, S. (2011) 'Improving an urban sustainability environment through community participation: the case of Emilia-Romagna region', *Procedia Engineering*, 21, pp. 1118–1123.

Puglielli, E. (2010) 'L'Aquila: le marginalità sociali dello shock, gli scenari dell'educazione', *Site.it Giornale*. Available at: www.site.it/le_testate/IL%20MARTELLO%20DEL%20 FUCINO/Puglielli%20-%20educare%20nel%20cratere1%20pdf.pdf (accessed 27 August 2014).

Regione Emilia Romagna (2009) *Partecipare e decidere. Insieme è meglio*. Regione Emilia Romagna: Quaderni della Partecipazione. Available at: http://partecipazione.regione. emilia-romagna.it/entra-in-regione/documenti/altri-documenti/partecipazione-1 (accessed 27 August 2014).

Regione Emilia Romagna (2014) *A due anni dal sisma*. Regione Emilia Romagna. Available at: www.regione.emilia-romagna.it/terremoto/a-due-anni-dal-sisma/a-due-anni-dal-sisma (accessed 23 September 2014).

Sbilanciamoci! (2010) *Rapporto QUARS 2010 indice di qualità regionale dello sviluppo*. Rome: Sbilanciamoci! Available at: www.Sbilanciamoci.org/docs/quars_2010.pdf (accessed 22 September 2014).

Tortoioli, L. (2014) 'La normativa per la ricostruzione e la politica dei centri storici', *Inforum*, 45, pp. 4–7.

Zanelli, M. (2014) 'Per una nuova cultura del paesaggio: dalla tutela alla valorizzazione', *Inforum*, 45, pp. 16–18.

9

FACTS ON THE GROUND

Affirming community identity through placemaking projects in West Bank villages

Jenny Donovan

Introduction

The West Bank is inhabited by two resourceful and resilient communities who feel a deep connection to the land, but share little else. For the Palestinians, this connection is derived from continuous presence on the land over many generations. It is not unusual for them to say they "belong to the land" as much as own it. However, to the Israeli settlers, nearly all of whom have come to the territory in the last 20 years, it is held equally dear as part of "greater Israel", a concept with historical, security and biblical foundations, and as a setting for ongoing community life. In recent years, the asymmetric balance of power that exists between the two peoples has meant that while Israeli communities have been able to plan for and realise a future that expresses their identity and meets their needs, the Palestinians have been denied the opportunity to do likewise (Hague *et al.*, 2015). This chapter seeks to examine a placemaking initiative by UN-Habitat that I was involved with in 2014–2015 that sought to partly address this issue. It did this by seeking to give Palestinian communities real-world, achievable, entry-level experience of planning to improve their conditions and meet their own needs.

Background

The West Bank has been occupied by Israel since 1967 after a failed invasion of Israel by neighbouring countries. Legally it is "under temporary belligerent occupation" (Shalev and Cohen-Lifshitz, 2008, p. 8). As examined below, this occupation has been accompanied by a skewing of planning and development rights against Palestinian communities while facilitating development by Israelis for settlement and military purposes (Hague *et al.*, 2015). In 1993, the Israeli government and the Palestine Liberation Organisation reached an agreement known as the Oslo Accords that

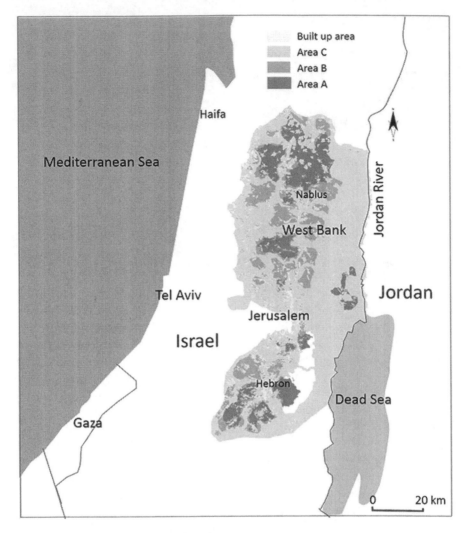

FIGURE 9.1 Map of the West Bank and Areas A–C.

Source: Map compiled partly from information supplied by UN Office for the Coordination of Humanitarian Affairs.

sought to overcome the ongoing humanitarian crisis brought about by the conflict and achieve a lasting resolution. This agreement envisaged an increasing level of Palestinian autonomy and divided the Occupied West Bank into a temporary (but at time of writing, still in place) patchwork of three different types of area (see Figure 9.1):

- Area A comprises the largest cities, within which the Palestinian Authority (PA) (representing ethnic Arab interests) has civil and security control.
- Area B comprises the more significant towns and villages. Within these areas, the PA has civil control and the Israeli government has security control.

- Area C comprises most of the rural areas and smaller villages. Within these areas, the Israeli government has security control, while the PA is responsible for some civil matters, such as education and health – a burden for the cash-strapped PA.

While Areas A and B account for the vast majority of the Palestinian population, they are discontinuous and together only cover approximately 40 per cent of the area of the West Bank. Area C accounts for the remaining 60 per cent (Shalev and Cohen-Lifshitz, 2008). The towns and villages scattered through Area C's arid landscape are of two very different types. On the one hand, there are the older, established Palestinian communities; on the other, there are the newer Israeli settlements. The Palestinian communities have a combined population of 300,000 (UN-Habitat, 2015). The Israeli settlements have appeared from the late 1970s onwards, and have expanded at a rapid rate. In 2013, there were 225 settlements and outposts and they were home to a combined population of 325,000 (Central Bureau of Statistics (Israel), 2012).

Planning as a means of control

Bimkom (an Israeli NGO that seeks to ensure planning and development decisions and systems reflect a commitment to human rights) finds that the Palestinian inhabitants of Area C do not enjoy many of the rights that the settlers enjoy (Shalev and Cohen-Lifshitz, 2008). Instead they suffer "violations of the right to life, security and bodily integrity, as well as violations of property rights", according to Israeli human rights group Yesh Din (2015, p. 5). Palestinians are subject to an ossified planning regime originally drawn up during the British mandate era in the 1940s. This is overlaid with ad hoc military rules and regulations that emphasise Israeli security over and above almost every other consideration (Shalev and Cohen-Lifshitz, 2008). One expression of this regime has been that the Israeli Civil Administration (the arm of the Israeli army that rules over planning matters in Area C) has complete control over planning, development and the provision of building permits (Shalev and Cohen-Lifshitz, 2008) and has refused most applications for development in Palestinian communities. Between January 2000 and September 2007, the Civil Administration approved just 91 applications for development – an average of 13 a year, or just 5.6 per cent of all applications. Without being able to build *legally* to meet the needs of the community, such development that has occurred in recent years has been technically illegal and so is under threat of demolition. During the same period 4,820 demolition orders were issued – an average of 714 each year – and 1,626 buildings were actually demolished, an average of 240 buildings a year (Shalev and Cohen-Lifshitz, 2008).

This ruthless exercise of power has eroded the ability of Palestinians to meet their own needs. Shalev and Cohen-Lifshitz (2008) found that most of the Palestinian residents of Area C have no regular access to water, electricity, sewerage or roads, and there is a distinct lack of public services such as schools, health clinics and employment zones. They further found that "even the construction of agricultural infrastructure is prohibited in most cases by the Israeli Civil Administration" (ibid.).

Another symptom of this highly weighted balance of power is the poor physical environment that blights Palestinian communities when contrasted with their Israeli equivalents. Such spaces offer people little in the way of opportunities and experiences to meet needs or express their identity and are constant reminders of their inability to exercise self-determination.

Facts on the ground

"Facts on the ground" is a phrase in common usage in the context of the West Bank. These are physical interventions that are intended to solidify de facto possession with the aim of attaining de jure possession and are based on the law of adverse possession. "It is a means by which a territorial base can be established and property converted into sovereignty" while wresting sovereignty from another group (Stollzenberg, 2009). Shalev and Cohen-Lifshitz (2008, p. 8) found "establishing facts on the ground through construction is the best way to secure effective control in the field".

In the West Bank, the most significant such "facts" are the Israeli settlements. They have been built in violation of international law as outlined in the Fourth

FIGURE 9.2 A typical street scene in the village of Bruqin, illustrating demolished buildings and poor public spaces characteristic of many villages in Area C.

Geneva Convention and are protected by "pragmatic power politics" (Stollzenberg, 2009). The UN General Assembly (2016, p. 2) noted "the presence and continued development of Israeli settlements lie at the root of a broad spectrum of human rights violations in the West Bank". According to the IAB, they are an "ever increasing obstacle to achieving a peaceful solution to the conflict" (Hague et al., 2015, p. 12). These settlements, together with other "facts on the ground", such as the separation wall and other aspects of the matrix of control, diminish options for potential political solutions and erode hope (UN-Habitat, 2015).

In another context, there would be much to admire in the design of these settlements. Typically located on hilltops, they enjoy commanding views. Each one incorporates a high level of social and utility infrastructure, they reflect meticulous design standards and typically include a high standard of landscaping. They are often surrounded by well-tended agricultural fields. They express the increasingly permanent, better-resourced, well-lit, well-serviced and powerfully defended presence of their inhabitants, compared to nearby Palestinian communities. These qualities help attract settlers (Lein, 2002) and allow them to make/reform connections to the land, accumulate memories and accrete experiences to a place and so embed themselves into the landscape. Thus these settlements "normalise" and legitimise the occupation by creating settled, maturing communities. They acquire moral weight as the settings for ongoing community life as settlers are born, live and are buried there, and invest a great deal of emotional capital into that land.

The rapid expansion of these settlements has made a significant impact, not just on the land but on the psyches of both communities. The 'facts on the ground' bring about not just the "physical facts of geography and possession of a particular piece of land, but also the myriad psychological and social effects unleashed by these facts" and bring about "profound alterations and transformations in the fabric of social existence" (Lein, 2002, p. 1).

For many Israelis, this momentum, projected forward, will cement their claim on this land and help them to realise their "manifest destiny", but for Palestinians (and indeed many Israelis) the settlements represent an appropriation that is profoundly unjust (Yesh Din, 2015). Not surprisingly, for many Palestinians, this creates a sense that they are being torn from the land and are being denied their rights and responsibilities to tend the olive groves, citrus groves, grapevines, etc. that have often been in their families for generations. This loss has been profoundly distressing. It is nothing short of a continuing, ongoing disaster and cultivates a sense that they are being disconnected from their land and consigned to lead diminished lives (Yesh Din, 2015).

Consequently, the outcome of assertive self-expression for one community and its denial to the other equates to a zero-sum equation, where the embedding of an Israeli presence has been paralleled by a corresponding erosion of a Palestinian one (Halper, 2000). Most of the actions that have caused this situation can be seen as resulting from an understandable desire for security. However, these actions diminish the rights of Palestinians and their ability to meet their needs, creating a tension between order and justice.

Local Outline Plans (LOPs)

Following a landmark legal case in the Israeli High Court in 2011, a mechanism was created to give Palestinian villages in the West Bank partial power to draft LOPs for the quantitative aspects of development such as the distribution of land uses and alignment of roads (UN-Habitat, 2015). These plans offer a degree of protection for these communities from demolition orders (ibid.) and allow some room for expansion.

UN-Habitat was appointed by the Government of Palestine to provide advocacy and planning support to the Palestinian communities in Area C. This programme sought to empower the Palestinians to visualise, plan for and achieve improvements to meet their needs. One strand of UN-Habitat's support was to provide advice on improving the qualitative aspects of development that became possible through LOPs. This was to be achieved through a process of "learning by doing" with a team of local and international UN staff and an overseas adviser. I was fortunate enough to be appointed to undertake this role. My "mission" with UN-Habitat (to use UN terminology) was to look at 14 villages throughout Area C and lead placemaking projects that sought to achieve the following:

- To promote a liveable place and introduce a unique identity for each village.
- To promote friendly and suitable aesthetics and the appearance of community.
- To encourage new patterns of designing, and protect existing natural and cultural heritage and historical sites and promote those sites' potential for tourism development.
- To promote proper investment in the community, and encourage local and eco-tourism.
- To facilitate a bottom-up process of design and implementation.
- To present people with what the professional considers a better place and innovative design.
- To study and interpret the local conditions.
- To balance users' needs and desires with consideration of design, engineering, costs and regulations.

(*UN-Habitat, 2014a*)

Placemaking

Placemaking is about imbuing a spatial entity (a space) with positive meaning for the people who share it and in that way make it a place. It seeks to change the relationship between people and their surroundings, so they may enjoy the experiences and opportunities that enable them to meet their needs, express their communal identity and fulfil their aspirations. Placemaking looks at the "software" of the place (the way people think about their surroundings and their sense of control over them) as much as "hardware" (what is built). Put another way, making "places" provides people with opportunities to interact with each other and with their shared surroundings to meet their needs and forge and strengthen social bonds (UN-Habitat, 2014b).

FIGURE 9.3 Example of a placemaking intervention.

Source: Copyright UN-Habitat.

Approach

This approach responds to the innate desire that people have to improve their shared surroundings and look after family and friends (Toker, 2012). It seeks to understand, respect and nurture these assets and translate them into shared aspirations for achievable development. The designer's role is to collaborate with the local community and arrive at an agenda for design that the local people can see reflects their aspirations and reconciles these with broader responsibilities. These include strategic objectives, the requirements of funding agencies and facilitating the community to maintain interventions in the long run. To this end, the placemaking process seeks to build a "community of implementation" of local people and stakeholders, for whom the achievement of the project matters. Because it is *their* project, they feel a sense of ownership. This emphasis on emotional capital allows the process to minimise dependence on limited financial capital to realise plans, but instead gives greater weight to a local community's assets and its self-determined identity.

The placemaking plans were envisaged to give these communities real-world "entry-level" experience of choosing, designing and implementing projects (with outside experts' help) to make their communities better suited to their needs and the challenges of the future.

For this project to have the optimal effect, I saw my role as being:

1   to provide the benefits of "real-world" experience of how the agenda estab-
     lished by a community might be translated into physical interventions that
     inspire, excite and empower locals to make changes to their surroundings and
     are appealing for donors; and
2   to give people ownership of the process by making it transparent, adaptable
     and replicable; hence it was broken down into "bite-size" steps that clearly
     showed how each step led to the next.

For this to happen, I applied a methodology adapted from past placemaking projects that had proven effective (UN-Habitat, 2014b), which was amended in

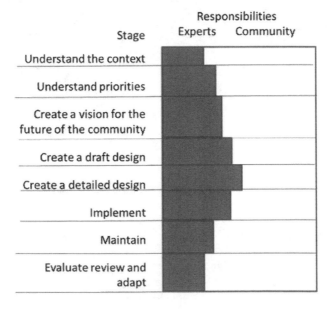

FIGURE 9.4 Changing balance of influence between community and designers.

Source: Copyright UN-Habitat.

consultation with local staff. This process sought to share responsibility between the community and outside experts (Palestinian and international) in a way that empowered the community while ensuring high standards of planning and design.

Engaging the community

In a land with patchy internet coverage and a virtually non-existent postal service, engagement occurred almost entirely through a series of workshop meetings. In keeping with local tradition, these were organised through the village leader, typically requiring many phone calls and often meetings where my colleagues outlined the process, its purpose and its aims. Getting traction was not without its difficulties: it is fair to say that placemaking, while resonating with many, was not seen as a priority, and was seen by some as a distraction. Our principal response was to emphasise that it complemented expected major engineering works and was facilitated by them, rather than replacing them. In this way, my colleagues created enough interest through patience and diplomacy to get 14 village leaders to commit to placemaking projects in their villages.

Each workshop was two-way. We communicated the process and how it might lead to changes and the community provided us with their insights and opinions. Each workshop was broken down into activities that were facilitated by *local* UN staff, in order that it might be interpreted as a local project backed up by international insights, rather than an initiative imposed by outsiders. Responses were invited at each session. Anonymous feedback forms and other parallel mechanisms made it easier for a wide section of the community to participate and for community members to hear others' views. At each subsequent session, we outlined our responses to each comment (again taking measures to secure anonymity) and invited people to comment or respond if we had left anything out. In this way, we sought to demonstrate that we had listened and it was the community who had control. No step was passed without community support.

Placemaking process

The first session in each village sought to ensure participants understood the project and the importance of their participation. We made an explicit commitment that we hoped would help cultivate trust. We sought to understand the social landscape; the spatial distribution of values, likes, dislikes, hopes and fears. This allowed us to draw conclusions about the direction and key goals that the project might pursue and identify or confirm candidate sites within the villages that may be most appropriate to address community concerns. This provided the 'community design agenda' that we drafted in a presentation drawing after the session. In a second workshop session, we explained what we had heard and the conclusions we had drawn, and the community was invited to confirm or amend them, again through multiple modes.

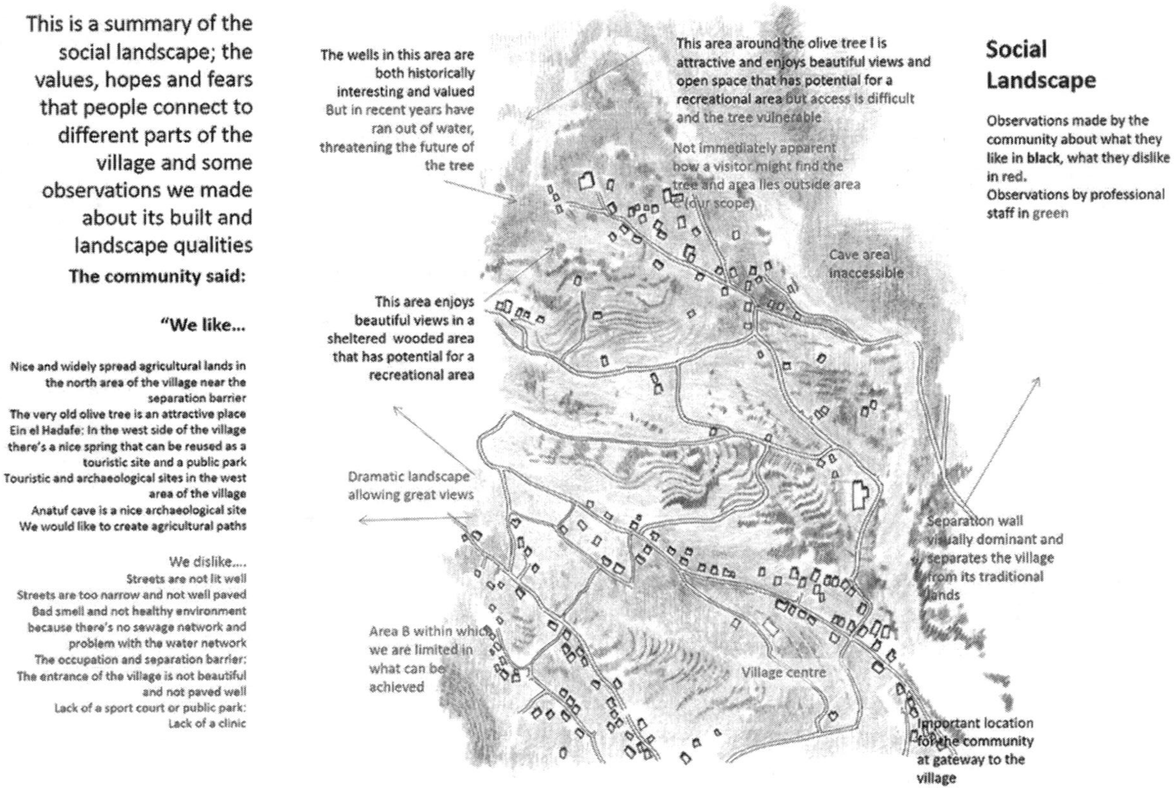

This is a summary of the social landscape; the values, hopes and fears that people connect to different parts of the village and some observations we made about its built and landscape qualities

The community said:

"We like...

Nice and widely spread agricultural lands in the north area of the village near the separation barrier

The very old olive tree is an attractive place
Ein el Hadafe: In the west side of the village there's a nice spring that can be reused as a touristic site and a public park
Touristic and archaeological sites in the west area of the village

Anatuf cave is a nice archaeological site
We would like to create agricultural paths

We dislike....

Streets are not lit well
Streets are too narrow and not well paved
Bad smell and not healthy environment because there's no sewage network and problem with the water network
The occupation and separation barrier:
The entrance of the village is not beautiful and not paved well
Lack of a sport court or public park:
Lack of a clinic

The wells in this area are both historically interesting and valued But in recent years have ran out of water, threatening the future of the tree

This area enjoys beautiful views in a sheltered wooded area that has potential for a recreational area

Dramatic landscape allowing great views

Area B within which we are limited in what can be achieved

This area around the olive tree I is attractive and enjoys beautiful views and open space that has potential for a recreational area but access is difficult and the tree vulnerable

Not immediately apparent how a visitor might find the tree and area lies outside area (A microscope)

Cave area inaccessible

Separation wall visually dominant and separates the village from its traditional lands

Important location for the community at gateway to the village

Social Landscape

Observations made by the community about what they like in black, what they dislike in red.
Observations by professional staff in green

FIGURE 9.5 Example social landscape (detail).
Source: Copyright UN-Habitat.

Once the direction was confirmed, the ideas were taken away and developed, then presented at a third workshop for discussion and explanation. The emerging concepts explicitly responded to the social landscape and design agenda and were designed to be implementable with local skills and also meet the agendas of the funding agencies. Again responses were sought through multiple means (presentations, discussions in the community and via email, design activities/games with the community and component parts of the community, response forms, via local UN staff and local champions), to assess how well the ideas resonated with the community. If required, a fourth workshop was undertaken with amended plans to ensure a general sense of satisfaction with the concept plans. Finally, these plans were documented and again presented to the community to establish and confirm support and start a conversation with collaborators.

The plans sought to reconcile planned major infrastructure projects with placemaking proposals by ensuring that the required rebuilding and reinstatement works could be utilised to create community assets and engage local community members.

The plans also recognised that many of the issues raised by the community fell outside the essentially small scale and local scope of locally implementable placemaking proposals, instead requiring infrastructure/engineering improvements. Rather than ignore the community-raised issues, the placemaking proposals were divided into two sections. Part 1 was a set of proposals for interventions and Part 2 was a list of other issues and their rationales that might help set the agendas of donors and other funding agencies.

The final designs used a palette of locally sourced materials and utilised local skills to ensure the designs could be built locally and reflected the *genus animi* (the spirit of the place). These plans were then compiled into attractively presented documents in English and Arabic that explained not just what was required but why it was required. This ensured that the areas of interest of funders and the local community were addressed. The plans also outlined the process and findings of the community engagement process, which were then used as a record of engagement and a marketing device to compete for funds and influence decision makers such as the PA and aid agencies.

Resonance with strategic objectives

A key aim of the project was to share awareness that a simple plan in the form of a shopping list of the communities' aspirations would serve little purpose. If it didn't meet strategic objectives of the Palestinian Authority or resonate with the agendas of key funding agencies (usually overseas aid organisations), it was unlikely to affect real change. To this end, an analysis of the priorities of these bodies was undertaken and a statement of stakeholders' shared priorities was drawn up to help guide the development of the placemaking plans. This statement was compiled from written declarations by the funding agencies and from critical analysis of the agenda of the Palestinian Authority by local UN colleagues. Placemaking projects helped in a number of ways (outlined below).

State-building and empowerment

The interventions outlined in the plans seek to provide people with real-world experience in taking control of their own surroundings, giving them opportunities to exercise a degree of self-determination. The projects were undertaken in a way that made the process accessible and allowed for the transfer of knowledge. Placemaking does not always require changing the legal use of a place, merely its qualities and the variety of activities it supports. Therefore, as long as it enjoys the legal protection offered by being within the boundary of the Palestinian outline plans, it has a relatively low threshold of permissions needed.

As such the placemaking projects sought to facilitate the prompt achievement of small but real "facts on the ground", expressing the unique character and inherent dignity of the communities they are embedded within. Experience from elsewhere in the world suggests such interventions provide a discernible sense that things can get better, helping keep the flame of hope alive. In my experience, this hope is a critical starting point without which envisaging positive change becomes almost impossible.

Support wealth creation

These interventions further seek to identify and unlock latent economic resources held within the communities. The workshops revealed these assets and gave the communities transparent experience of developing and testing ideas to utilise them. By realising them through community contracting their implementation supports local business and assists people to exercise and develop local skills. It also offers communities the 'know how' of working with professionals to develop and utilise skills. This will make the most of the unique character of each community and can help increase visitation, potential employment opportunities and the economic vitality of these communities.

Facilitate sustainable development

These projects were intended to allow people to enjoy surroundings that are conducive to meeting their needs as well as facilitating households and communities to thrive and fulfil their potential with minimal dependence on outside resources. The interventions emphasised the efficient use of local resources and sought to ensure all interventions were designed to meet multiple local needs. The combination of outside experts and a local agenda allowed international best practice to be considered and applied at a local level.

Democratic governance and human rights

The placemaking process is intended to provide an achievable, real-world model in which interventions in people's shared surroundings are transparent and respect a community's right to self-determination. The dialogue that placemaking makes possible between experts and community also ensures that people can make well-informed choices and that local rights are not at upheld the expense of the broader community or Palestinian society as a whole. Furthermore, it provides a way of validating and recognising community views and aspirations and demonstrates a level of esteem placed on the relevant community.

Outcomes

The plans and the methodology behind them won funds for implementation by the EU in the first four communities. These have now been constructed and their impact will be assessed post-completion. Initial feedback (Johannes, 2015; D'hondt, 2015) suggests the following:

- The plans have enabled some people to look upon their surroundings as catalysts for hope. They offer a way of expressing their own values and cultural identity and affirming their ongoing connection to the land.
- They have provided an accessible model for communities to "cut their teeth" on making decisions and coordinating work. Anecdotally, my colleagues are hearing that other communities (other than the initial four) have independently sought funding for their projects.
- The persuasive and clear articulation of the ideas, the values they were based on, and the process by which they were defined have raised interest with potential donors.
- While these Palestinian "facts on the ground" are vastly asymmetric with Israel's in terms of grandeur, their legacy is not just physical. They have added to the resources at hand to enable people to stake their claim on their land and accrue the experience to make bigger interventions.

However:

- While we requested the attendance of a cross-section of the community – men, women, children and older people – at the meetings, we didn't always achieve this. Instead, we often got only the village leader's inner circle, who were mainly men, which raised questions about representativeness.
- Although hopes were cultivated in many communities, the articulation of concepts before a budget is confirmed may lead to those hopes being dashed and makes people wary that the process may not come to anything. For many, this will just reinforce a sense that consultation can be meaningless.
- We are still waiting to see if the project has provoked locally inspired indigenous placemaking projects.

Observations

The project was undertaken against a very complex and dynamic backdrop. Apart from the agenda of the Israeli authorities and settlers, as touched upon above, there were many agencies, local and international, who were pursuing a variety of agendas in Area C. Reconciling our work with these agencies and the priorities of the local communities added another level of complexity and would not have been possible without my local colleagues' skills in diplomacy and persuasion.

The project confirmed the strength of the connection that people have to their community and the tenacity with which they hold on to it. However, we only had access to those in the community who attended the meetings. As noted, these were typically community members connected with the village leader. We were aware that there were probably other perspectives, but recognised that to seek them too forcefully would risk causing conflict. Hence, we took the pragmatic view that any extension of self-determination through a community was better than none.

It also revealed the uneven balance of hard and soft power in Palestine. Hard power was evident from the walls and barriers that snake across the landscape and the settlements atop many hills. Soft power was evident by the way those settlements influenced people's thinking about what represented success. On more than one occasion, when I asked a Palestinian what represented good design, they pointed out of their window to the orderliness, consistently pitched roofs and landscaping of a neighbouring Israeli settlement rather than the examples of contemporary architecture that exist in their own community.

We noted the importance of gender perspectives. As is common in Arabic cultures, men and women would form into single-sex groups in the workshops to discuss ideas and often had very different priorities. To equalise these contributions, we invited each table to report back their findings, not as the last word but as their perspective, which helped ensure that the perspectives of both genders were given equal weight.

The LOPs provided the statutory underpinnings for the project and were the main source of base data. However, in several cases many in the communities felt aggrieved by the provisions of these outline plans and these grievances spilled over into discussions about the placemaking plans. These distracted participants and compromised our ability to gather the information we needed.

We also noted that attendance at meetings varied between workshops. When each step was supposed to progress from the previous step, this created a dilemma. To repeat the introduction and purpose at every step, ensuring everyone understood the role and context of the workshop, risked repetition for those who had been there before. However, not to repeat this material could leave people confused or labouring under misconceptions.

In several of the communities, 'consultation fatigue' was evident; people had been consulted many times by a plethora of different agencies. They had often been asked apparently similar questions at each of these meetings, but nothing (or very little) seemed to be achieved. Hopes were often raised only to be dashed.

We further noted some participants in some communities were predisposed against the project because the idea of placemaking seemed alien to them, an imposition. Past experience of consultation exercises, confusion and inaccurate assumptions about the scope of the project also created reluctance. In other communities, people were inclined to say what they thought we wanted to hear, responding "Yes, thank you" to whatever they perceived we were offering and not wanting to 'rock the boat'. In yet other places, the LOPs were perceived as collaborationist (D'hondt, 2015) and the placemaking projects little more than bribes.

Conclusions

This initiative suggests that placemaking projects can be catalysts for hope and provide "entry-level" community development opportunities. However, reflection on this initiative suggests that future projects may benefit from the following:

- More obvious leadership by local placemakers: this was always intended for future placemaking projects, but perhaps could have been made clearer.
- Clarification of the role of placemaking: in order to avoid the recurring suspicion that placemaking was being 'pushed' onto them as a substitute for more basic infrastructure, such as water and electricity.
- Greater detail: originally the drawings in the proposals were primarily in the form of plans and perspective sketches that were deliberately "detail sparse", to make sure they could be understood by laypeople and because experience suggests these are good at capturing the attention of funders. Furthermore, such graphics effectively capture the feeling of a scheme and allow flexibility in interpretation. This creates a model that local people could adapt to their own needs rather than constrain them with documented details that reflect a more technocratic approach. However, given that the experience of implementing these projects suggests that contractors will be needed for at least some tasks in the future, it is recognised that more detailed specification drawings are necessary to ensure that what is built suits its intended purpose.

Furthermore, my colleagues suggested that to optimise the contribution of placemaking projects, they should be coordinated at the regional/strategic level and incorporated into "placemaking activism". These could be short, intense periods of design and immediate implementation led by locals with technical advice supplied by outside experts.

> While doing the designs for a better place we could include a component to mobilise local forces to clean up the village, do some acupunctural interventions such as tree and flower planting, temporary bins, cultural placemaking events, youth/women mobilising events, temporary street interventions to calm traffic or mark a future public space etc.
>
> (D'hondt, 2015)

References

Central Bureau of Statistics (Israel) (2012) *Israel Statistical Abstract 2012, Table 2.6, Population by District, Sub-District, and Religion.* Available at: www.cbs.gov.il/shnaton63/st02_06x. pdf.

D'hondt, Frank (2015) Personal correspondence with the author.

Hague, C., Crookston, M., Gladki, J., Platt, C. and Wegener, M. (2015) *Spatial Planning in Area C of the Israeli Occupied West Bank of the Palestinian Territory.* Report of an International Advisory Board. Jerusalem: UN-Habitat.

Halper, Jeff (2000) *The 94 Percent Solution.* Middle East Research and Information Project. Available at: www.merip.org/mer/mer216/94-percent-solution (accessed January 2016).

Johannes, Judith (2015) Personal correspondence with the author.

Lein, Yehezkel (2002) *Land Grab: Israel's Settlement Policy in the West Bank.* Report for B'tselem. Available at: www.btselem.org/download/200205_land_grab_eng.pdf (accessed December 2014).

Shalev, Nir and Cohen-Lifshitz, Alon (2008) *The Prohibited Zone: Israeli Planning Policy in the Palestinian Villages in Area C.* Report for Bimkom. Available at: http://bimkom.org/ eng/wp-content/uploads/ProhibitedZone.pdf (accessed December 2014).

Stollzenberg, Nomi (2009) *Facts on the Ground.* University of Southern California Law School, Legal Studies Working Paper Series, no. 45. Available at: http://law.bepress. com/cgi/viewcontent.cgi?article=1125&context=usclwps-lss.

Toker, Umut (2012) *Making Community Design Work: A Guide for Planners.* Washington, DC: American Planning Association Planners Press.

UN General Assembly (2016) *Israeli Settlements in the Occupied Palestinian Territory, Including East Jerusalem, and the Occupied Syrian Golan.* Report of the Secretary-General. New York: UNGA.

UN-Habitat (2014a) Placemaking Expert International Consultant Mission Terms of Reference (unpublished consultancy brief).

UN-Habitat (2014b) *Public Space in the Global Agenda for Sustainable Urban Development: The 'Global Public Space Toolkit'.* Available at: www.urbangateway.org/sites/default/ugfiles/ Global_Toolkit_for_Public_Space.pdf (accessed January 2016).

UN-Habitat (2015) *'One UN' Approach to Spatial Planning in 'Area C' of the Occupied West Bank.* Available at: http://unhabitat.org/wp-content/uploads/2015/10/One-UN-Approach-to-Spatial-Planning-in-Area-C-.pdf.

Yesh Din (2015) *Under the Radar: Israel's Silent Policy of Transforming Unauthorized Outposts into Official Settlements.* Available at: www.yesh-din.org/userfiles/Yesh%20Din_Under% 20The%20Radar%20-%20English_WEB%283%29.pdf.

10

WOMEN AND THEIR ROLES IN PEACE BUILDING IN CONFLICT-VULNERABLE AREAS OF MINDANAO, PHILIPPINES

Beau B. Beza, Mary Johnson and Anne Shangrila Y. Fuentes

Introduction

The Philippines is a disaster-vulnerable country experiencing typhoons, earthquakes, volcanic eruptions, landslides and climate change related issues (i.e. drought). Yet, for over four decades, the Mindanao region of the Philippines has experienced another form of disaster—violent armed conflict. This protracted conflict situation in Mindanao is complex, involves multiple parties (i.e. the Philippine Government, separatists, communists, clan militias and criminal groups) and has resulted in high rates of poverty and displacement. For example, since January 2015, close to 220,000 people have been displaced in Mindanao due to conflict and violence (Internal Displacement Monitoring Centre, 2016).

For the Philippine Government, the development of peace and security in Mindanao is a major priority, emphasised in the Philippine Development Plan 2011–2016 and also by the proposed Bangsamoro Basic Law (BBL) bill, which is a product of the ongoing peace talks between the Philippine Government and the Moro Islamic Liberation Front (MILF). A significant factor in the Mindanao conflict is income deprivation, along with social dislocation and isolation from services. The provinces within the Autonomous Region of Muslim Mindanao (ARMM) and neighbouring areas (see Figure 10.1) remain at the bottom of the scale within the Philippines in terms of real per capita income.

Since 2013, Australian and Philippine research teams have been jointly working with conflict-vulnerable communities in Mindanao to co-produce, test and evaluate extension methods developed to achieve community-expressed outcomes. Their research, titled the Australian Centre for International Agricultural Research Mindanao Agricultural Extension Project (AMAEP), specifically investigates how community-based agricultural extension methods, developed in conflict-vulnerable areas, can enhance villagers' livelihoods. The ongoing community engagement and

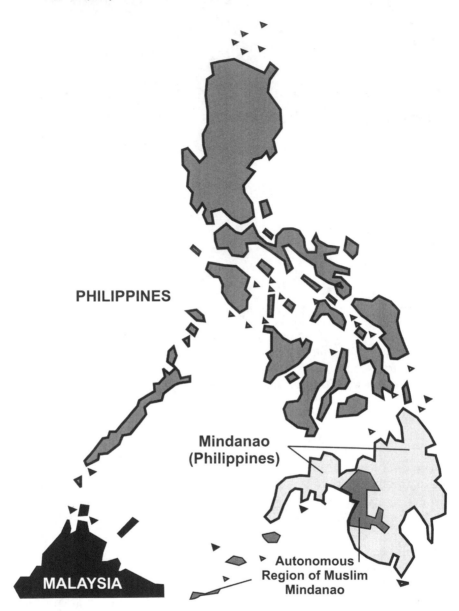

FIGURE 10.1 Map of Mindanao, Philippines, and the Autonomous Region of Muslim Mindanao.
Source: Adapted from Wikimedia.org and 2008 Philippines Map Destination 360.

project activities are delivered by locally based and trained facilitators. Having come from the respective study site areas, where the research activities are being conducted, the facilitators understand local and wider issues, have well-established networks and are trusted by the participating communities.

The overall aim of the AMAEP project is to develop an improved 'model' for agricultural extension in conflict-vulnerable areas of Mindanao and has four objectives:

1   Determine the livelihood impacts of conflict on agricultural communities and extension services in case study conflict-affected areas.
2   Implement a pilot programme of improved extension and livelihood innovations, making use of principles and methods largely derived from previous ACIAR projects.
3   Analyse the impacts of the pilot extension and livelihood innovation programme.
4   Engage more broadly with relevant conflict-area extension and other agencies outside the case study areas to communicate project methodologies and findings.

The AMAEP project utilises a three-track approach to meet its aim and objectives. That is, it focuses on improving farmers' access to technical innovations, builds community social capital and collaborates closely with local institutional partners. This three-track approach recognises lessons learned from previous projects, where technology was the sole focus and consequently had less successful results.

This type of AMAEP in-field livelihood improvement effort in conflict-vulnerable areas is uncommon. However, at the national government level a number of larger policy commitments exist. For example, the Australia–Philippines Development Cooperation Program Statement of Commitment 2012–17 (referred to as the AusAID Philippines Country Strategy) is directed by two strategic objectives: 1) strengthening basic services for the poor; and 2) reducing vulnerabilities from climate change and conflict. In line with this, the Australian Centre for International Agricultural Research's (ACIAR; an Australian Commonwealth government agency) medium-term strategy for the Philippines encompasses technology uptake and extension processes in Mindanao, with specific reference to conflict zones.

Even less researched and reported is the role women play in peace-building efforts. And although not explicitly stated as a research object, the Australian/Philippine AMAEP team members have been captivated by the varied roles women play in the Mindanao peace-building process. This chapter reports on the peace-building role of women in the AMAEP project study site in the Ipil municipality.

Women and peace building

Conflict occurs in various forms in the Mindanao region, from local community disputes to *rido* (i.e. clan conflict) and, on a greater scale, clashes between armed opposition groups and the government. These conflict situations are complex and to describe them in terms of religion, culture or ethnicity is too narrow, and typifies the approach of many mainstream international development institutions, central government and some scholars (Vellema and Lara Jr, 2011).

The *Gender and Conflict in Mindanao* study (see Dwyer and Cagoco-Guiam, 2010) notes that while women are often disproportionately affected, the impact conflict has on women and girls is frequently misunderstood, manipulated or ignored. This 'mis-understanding of the impact on women' is reported in other conflict-vulnerable areas, such as in several South Asian countries (e.g. Manchanda, 2005). There are important differences between communities, as well as among individual women, that shape their experiences of conflict. At the same time, strong commonalities emerged across diverse constituencies and understanding the various divergent impacts of conflict on women is necessary for effective, empowered and engaged communities.

Studies show that conflict-related constraints to mobility are a major concern (Dwyer and Cagoco-Guiam, 2010; Jordan, 2003; Vellema and Lara Jr, 2011), which result in some degree of gender role transformation. Women are less tar-geted for revenge killings and remain comparably mobile. Consequently women take on activities and responsibilities, including income generation, traditionally performed by men. Dwyer and Cagoco-Guiam (2010) note that women in Min-danao tend to view their role as the breadwinner during conflict less as an example of empowerment and more as an exhausting strain. Moreover, when conflict flares, more rigid social norms regarding women's appropriate roles may become more central to identities, circumscribing women's choices and even creating divisions among women, as some women resist those norms and others embrace them. In terms of a male view of this change, Moser (2007) suggests that as women's roles evolve, men view this change as negative and refer to their loss of status and respect in a community. Greater recognition of women's varying roles in peace build-ing and community strengthening, post-armed conflict, is gaining momentum (see Jordan, 2003; Moola, 2006; Moser, 2007). Through their social contributions as educators, economic actors, mothers, community mediators and leaders of civil society groups, women hold crucial roles for shaping peace in conflict-vulnerable communities. Jordan (2003, p. 239) in her work on '[giving a] voice to women working in the field' of peace making organises the above roles into themes that help to frame women's responsibilities. These are:

- Supportive: enabling, assisting, facilitating, supporting, accompanying, and building up.
- Directive: organising, training, managing, advising, and providing resources and information.
- Networking: promoting, liaising, disseminating, publishing, and influencing.
- Representing: acquiring the roles of ambassadors and advocates.

(Jordan, 2003, p. 242)

Dwyer and Cagoco-Guiam (2010) also observe that Mindanaoan women do not see themselves as passive observers to conflict, and report high levels of wom-en's participation in small-scale dispute resolution training. This type of women's involvement is referred to by Hunt and Posa (2001) as 'inclusive security', which is a concept that regards women as change agents rather than passive participants or observers in peace making. Yet women face numerous challenges in their 'new roles' and are often presented with community fragmentation along religious, ideological and class lines, with some groups subordinating discussions of their core gender issues to claims of nationalist or religious identity.

Women also play a significant role in local (Mindanaoan) *bayanihan* activities—the communal tradition where everyone works together for the common good. The longest history of *bayanihan* is found in agricultural communities; however, the tradition has diffused throughout Filipino society and is an expression of team spirit and sharing of labour (Gibson *et al.*, 2010; Oracion *et al.*, 2005).

The role of Filipino women in peace building may be traced in the country's history. Historically, women in the Philippines have been actively involved in society, including peace-building efforts. Before the Spaniards colonised the Philippines, Filipino women acted as healers, advisers, mediators and *babaylans* (priestesses)—a highly respected role in the community (Quindoza-Santiago, 2007). Women fought as bravely as men during the 1896 revolution when the country struggled for independence against Spain. Women like Gabriela Silang and Gregoria de Jesus were among the revolutionary leaders at that time. This tradition carried on with the Suffragist Movement winning the right to vote in 1937 when women participated in a plebiscite. Women were actively involved in society during the American colonisation and after. Several women's organisations emerged, especially during the administration of the first Filipino woman President, Corazon Aquino. The Philippine women's movement is still a vibrant force today.

Women occupy strategic roles in the peace process

Women directly affected by conflict should have and do have the ability to be part of the solution for peace (Manchanda, 2005). However, effective participation must ensure that women's voices are integrated into all relevant decision-making processes.

One of these processes includes women playing prominent roles in both the formal and informal aspects of the peace negotiations. In a formal sense, there is the opportunity for women to participate in the Philippines-based Office of the Presidential Adviser on the Peace Process (OPAPP), headed by Secretary Teresita Quintos-Deles. The OPAPP is a cabinet group placed directly under the Office of the President and tasked with 'the coordination and implementation of all the components of the comprehensive peace process' (OPAPP, 2015). Until recently, peace talks with the Philippine Government, the MILF peace panels and support staff comprised all-male delegations. Then, in 2012, two women were included in the technical staff associated with these groups; one of these women, Attorney Raissa Jajurie, plays an active role at the negotiating table on peace-building matters. Significantly, the head of the government's peace panel is Professor Miriam Coronel-Ferrer, and she has been active in driving the peace process.

The strategic positions occupied by these two women reflect recognition of their capacity to lead in conflict resolution at the national level (OPAPP, 2015). Such a capacity can also be gleaned at the local level through equally capable women, where they are considered by their communities as communicators, mediators and negotiators. Importantly, women are empathetic to their communities and 'share the conviction that feelings and healing are of central importance to peace building' (Jordan, 2003, p. 249).

In contrast to these women's national-level peace building efforts, in Mindanao the greatest involvement of women in conflict resolution remains at the grassroots level, where they have played key roles in mediating long-running conflicts (Pennington, 2012). However, and in addition to this mediation role, women have held major positions in evacuation centre management, assisting in the rationalised distribution of relief, and performing other community roles, bridging the evacuees or internally displaced persons (IDPs) with external donors and support groups. In Datu Piang (a municipality in the southern Philippines), several women have already shown their leadership qualities during relief distribution in the evacuation centres, ensuring that relief operations become orderly. Yet, these same women feel frustrated that their leadership potential is not recognised (Dwyer and Cagoco-Guiam, 2010).

Collection of data

In the first year of the project, the AMAEP utilised seven methods to measure stocks of social capital and build an overview of the participating farming communities. Of the seven methods, five gathered data relating to the role of women in peace keeping:

1   A baseline survey

    •   farmer
    •   institutional, agency and *barangay.*[1]

2   Community and institutional mapping.
3   A social capital review within the farmer groups ('group health').
4   Case studies.
5   Project staff and other key informant observations (i.e. local government units and *barangay* officials).

The use of multiple sources reflects the multi-dimensional nature of social capital and highlights that one or two sources alone are inadequate for developing a clear situational understanding.

Baseline data plays an important role in developing, confirming or negating the accuracy of a situation analysis. Two baseline surveys were conducted at the commencement of the project: 1) a farmer survey; and 2) an institutional, agency and *barangay* survey. The farmer baseline surveys were undertaken at each of the three initial pilot sites: *Barangay* Magdaup, municipality of Ipil, Zamboanga Sibugay; *Barangay* Assumption, municipality of Koronadal City, Province of South Cotabato; and *Barangay* Kauran, municipality of Ampatuan, Province of Maguindanao. A total of 185 one-on-one and group interviews were conducted by AMAEP facilitators using a set of pre-tested survey questions. The baseline data was then entered into SPSS software for analysis. Cebuano, Tagalog or local dialect responses were translated into English by project team members fluent in these dialects. In the following sections of this chapter, data from respondents interviewed in the study sites are reported and discussed. A case study using data generated from the Ipil municipality pilot site draws out further lessons below.

The institutional, agency and *barangay* surveys were conducted over March, April and May 2014 by the project team through one-on-one interviews with *barangay* officials, local government units (LGUs), extension agencies and other higher-level political entities (e.g. Moro Islamic Liberation Front (MILF), Bangsomoro Development Agency (BDA) and Mindanao Development Authority (MinDA)). A total of six surveys at the three pilot sites were conducted.

Results from the baseline survey

This section describes in detail results from the baseline surveys relating explicitly to women's roles in farming communities. Survey questions sought to uncover details of community networks and the primary forms of communication used between these networks and groups, the role and level of involvement of women (and children) in farm operations and in community groups, perceptions of the impact of conflict (what/how) on their farm business and social cohesion, in addition to the perceived impact on conflict/peace of improving farm business performance. To aid in the reporting of data, quotation marks are used to denote respondents' responses.

Research team members asked respondents to identify the household primary care giver, that is, the primary person who tends to the needs of the family and household. This was asked in order to gain an understanding of the roles and responsibilities of women and to identify any additional pressures during times of conflict and disruption. For example, in conflict-affected areas, male relatives typically avoid public spaces (i.e. farm fields, market places) and, consequently, important farm activities can be disrupted. Women in the study sites were found to take on these roles, which, coupled with their domestic obligations (e.g. cooking), placed extra pressure on them and their households. Nonetheless, taking on this role is dependent on the severity of the conflict, which, in high-conflict areas, also impacts on women's mobility.

In the Ipil municipal site, respondents reported mothers as primary carers in households and observed that this role has added demands during times of conflict. One respondent commented that there are 'more responsibilities at home since there are no classes'. That is, when conflict occurs children stay at home during the day rather than attend school and this necessitates parental supervision, which the primary carer(s) provide. Additionally, the Kauran *barangay* respondents reported that care giving was relatively balanced between fathers and mothers looking after children. However, these respondents also reported that during conflict, the pressure placed on families can be extreme; for example, one female respondent's house became the evacuation centre for indigenous people whose families were connected to her husband. Not surprisingly, conflict was observed to greatly increase levels of stress within communities and trigger income loss, food shortages and illness.

Farming is an important activity in all of the study site communities. Particularly, women's involvement in farming was described in terms of undertaking day-to-day farm tasks and (along with men) adopting new agricultural technologies and participating in training and field demonstrations. Additionally, women (as well as family members and neighbours) are called upon in response to peak seasonal activities such as sowing and harvesting. These reported activities highlight the importance that women play in terms of both farm labour and training, and also farm decision making.

For all three sites, women were reported as members of community groups, in addition to holding leadership positions in local organisations. However, there were differences in the ratio of women to men. For example, there were fewer women community group members than men in the Ipil municipality, while in the municipality of Ampatuan there were equal numbers of the two sexes in the groups, and the Koronadal municipal area had more women than men in the community organisations.

The farmer baseline survey also explored the impact of conflict on women. A caveat on these responses is that the majority of survey respondents are male, and further interview work is being undertaken to seek women's views.

Conflict had a significant effect on people's mobility, with survey respondents across all sites reporting that during this time people rarely go to neighbouring villages and children are too scared to go to school (hence the previous comment about children staying at home). Women are particularly vulnerable and, when displaced, respondents reported that women have problems with sourcing food and shelter, and attending to the health and education needs of their family members. Evacuation was particularly difficult for women with small children, and respondents 'experienced illness in evacuation sites'. Women also found it difficult to manage any work commitments or income support during times of conflict. For example, it was reported that 'the women were scared and they lacked food, they waited for government support as they had no money at all'.

In summary, women in the study sites were primary care givers, widely involved in farm work, contributed to household incomes, and were active in their local communities. However, during times of conflict and displacement, women were particularly vulnerable and burdened as they took on additional roles to help feed their families, and provided safety and shelter for a range of family members.

The role of women in peace building

Across all of the surveyed AMAEP sites, women were considered vital in performing a role towards peace. In particular, women's communication and negotiation strengths were noted by survey respondents, as was their ability to provide advice to their men and their families. For example, women are 'the ones who make their children aware of conflict happening'; they advise the kids 'not to anger Muslim brothers and not to generalise them as enemies'. They also taught children to respect *barangay* officials and advised the respective youth and men to 'stay cool-headed' so that conflict could be avoided. Women also encouraged their families and others to participate in activities in the *barangay*, especially those that initiate and conduct training to improve people's lives and, importantly, promote cooperation and peace building.

These communication skills came to the fore when women acted as mediators both within and outside their communities. The ability of women to listen was frequently cited in the context of understanding and empathy: 'they are known to have a soft and sympathetic heart', as one respondent explained. Women in the study sites were also described as community defenders: 'they tell the military to stay away from their community'.

The dissemination of reliable information is critical in times of conflict and women 'disseminate information to all people' in the community in order to raise awareness; they were also trusted and 'believed'. Women were also able to participate in decision making, since they have softer voices and can mediate conflict. They can help in providing food assistance to men who watch an area at night and take care of children while in evacuation sites.

Once conflict has subsided, women return to their daily activities, and farming women help to rebuild their communities in both social and economic terms. They encourage attendance at meetings and activities that promote cooperation and peace. During these meetings the women suggested 'building up love, peace and respect for other tribes'.

Ipil study

The municipality of Ipil has an area of 241.6 square kilometres and is the capital of Zamboanga Sibugay. It has experienced ongoing conflict for some years between the government and Abu Sayyaf, the government and MILF, and the government and the New People's Army. The AMAEP project identified *Barangay* Magdaup, one of 28 *barangays* in the municipality, as its study site. Magdaup is composed of a mixed group of Muslims, Christians and a range of indigenous people. This community resides 3 kilometres from the centre of Ipil.

In 1995, the municipality's town centre was the site of a bombing by the Abu Sayyaf group. At least 50 residents died when the town was attacked by armed men and the public market was set alight. People have lived in constant fear of violence erupting again. In October 2011, the nearby towns of Alicia, Kabasalan and Payao experienced violent clashes and armed conflict. In the same period, MILF and government soldiers clashed in Ipil resulting in numerous casualties. Recently, a retired education official and her grandson were abducted from *Barangay* Bangkerohan. Insecurity and threat remain in this town.

The AMAEP project facilitators began working with *Barangay* Magdaup, specifically in the *sitios*[2] of Sampaloc and Katipunan in 2014. Sampaloc consists primarily of tenant farmers who cultivate maize (mono crop). Over a number of seasons, the farmers had experienced poor-quality crops with low yields. Hence, they were keen to learn and increase their agricultural knowledge and skills. Katipunan, also within Ipil, is isolated with a large transient population. The facilitators reported that in this *sitio* there is significant malnutrition among children and high levels of mistrust among the community. For example, the LGU-MAO personnel were reluctant to enter the area because of its perceived security problems. In summary, the two *sitio* farming communities were very poor, with few support services or provision of information and high levels of stress and child malnutrition.

Initially, the AMAEP facilitators worked with the Magdaup farmers to assess their farms and identify their aspirations for improved livelihoods. Through that activity, vegetable gardening was identified as a priority project for improving

FIGURE 10.2 Members of the MVGA selling their produce.

farmers' livelihoods. As a result of this prioritisation, the farmers have participated in training and a series of activities including field schools, cross-visits, and hands-on technical practicals in vegetable production.

Significantly, the same farmers recently formed the Magdaup Vegetable Growers Association (MVGA). The association includes both women and men, and Muslims and Christians. Figure 10.2 shows a market stall where members of the MVGA were selling their produce. Interestingly, the AMAEP project facilitators observed that the relationship among the association's members has vastly improved over time. Where previously there seemed to be a lack of trust between Muslims and Christians, members are now keen to help each other. They not only visit each other's farms, but also provide advice on how to improve vegetable production and address problems on the farms. One Christian farmer association member remarked that it was her first time entering Katipunan (a Muslim area), even though she had been living in the region for a very long time. Furthermore, AMAEP facilitators are now linking the MVGA to local government and other agencies to create opportunities for participation in various programmes.

FIGURE 10.3 Christian and Muslim MVGA (women) members working together to plant seedlings.

This change in the relationship between Christians and Muslims reflects a growing level of trust between previously mutually suspicious groups. It is an interesting outcome of the project's ability to bridge and build trust along the way through the members' interactions as participants in the project's training and other activities.

Significant achievements for association members have been the adoption of new technologies, such as the introduction of container-gardening technology for crops. These vegetable-growing enterprises have done well and contribute to increasing levels of income generation, food security and nutrition. In addition, the association's women members provide vegetables for their children to sell at school for lunch money. Figure 10.3 shows Christian and Muslim association members planting seedlings together.

Conclusion

The AMAEP is developing a model of improved livelihood extension through facilitated community engagement with conflict-vulnerable communities. Gaining a better understanding of communities' social capital levels (i.e. networks, trust, social relationships) has also revealed the significant role that women play in recovery and peace building.

Three key conclusions can be drawn from the study results so far. The first is that women have significant roles as communicators, negotiators and advisers when working to improve prospects for peace. These roles are evident in the study site areas and lends support to the work of Dwyer and Cagoco-Guiam (2010), which identified the women of Mindanao as active participants in peace building. These roles can also be argued to 'fit' within Jordan's (2003) thematic framing of women's varied activities that help contribute to peace building. In particular, women's communication and negotiation strengths were noted by AMAEP survey respondents, as was the ability of women to provide advice to their husbands and families. Mindanaoan women, often unrecognised by formal structures, are recognised informally by their local communities as negotiators, mediators and advisers who aid conflict resolution.

A second conclusion drawn from the data relates to women's capacity to accomplish activities that were previously regarded as a male roles. In times of conflict, when men's mobility is restricted, women often take on the additional responsibilities of field work or travelling to the market to buy and sell produce (as well as tending to their domestic obligations and responsibilities). In essence, during conflict these women become custodians of the farm and household by providing non-threatening means of engaging with other community members and people external to their village setting.

A third conclusion, as described in the Ipil investigation, is that trust levels between Christian and Muslim women have increased via their participation in group activities. An important cultural/societal form of group work practised in many parts of the Philippines is the *bayanihan* system, where farmers' work groups make voluntary work contributions for communal benefit. This *bayanihan* system, previously not practised between Muslims and Christians, has now emerged in the Magdaup community. Importantly, the members of the MVGA, which is dominated by Christian and Muslim women, feel that collaboration is key to the association's success. This collective cooperation among women reflects growing trust due to frequent interactions and exchanges between community members.

Overall, the data gathered by the AMAEP project highlights the important and emerging roles of women in peace-building efforts. The women are considered proficient communicators and negotiators who are working towards overcoming mistrust as well as supporting community-based recovery processes. The respondents in the three study sites consistently noted women as key players in the peace-building process. Women's roles in this process are not recognised because of a lack of male influence, but rather because women are seen as contributing to the greater good of the village and the wider community.

Acknowledgement

This chapter originates from a paper titled 'Working Paper No. 19: Women's Roles in Peace Building', authored by Mary Johnson, Anne Shangrila Y. Fuentes and Beau B. Beza (December 2015). The working paper forms part of a series generated for an Australian Centre for International Agriculture Research (ACIAR) project titled 'Improving the Methods and Impacts of Agricultural Extension in Conflict Areas of Mindanao, Philippines (ASEM/2012/063)'. All of the working papers are available on the project website: https://sites.google.com/site/improved extensionproject/home.

Notes

1 A *barangay* is an administrative division in the Philippines—a village, district or ward.
2 *Sitios* are the smallest rural subdivision of a *barangay*.

References

Dwyer, L. and Cagoco-Guiam, R. (2010) *Gender and Conflict in Mindanao*. The Asia Foundation. Available at: http://asiafoundation.org/resources/pdfs/GenderConflictinMind anao.pdf.

Gibson, K., Cahill, A. and McKay, D. (2010) 'Rethinking the dynamics of rural transformation: Performing different development pathways in a Philippine municipality', *Transactions of the Institute of British Geographers*, 35(2), 237–255.

Hunt, S. and Posa, C. (2001) 'Women waging peace', *Foreign Policy*, 124, 38–47.

Internal Displacement Monitoring Centre (2016) *Philippines IDP Figures Analysis*. Available at: www.internal-displacement.org/south-and-south-east-asia/philippines/figures-analysis (accessed 7 June 2016).

Jordan, A. (2003) 'Women and conflict transformation: Influences, roles, and experiences', *Development in Practice*, 13(2–3), 239–252.

Manchanda, R. (2005) 'Women's agency in peace building: Gender relations in post-conflict reconstruction', *Economic and Political Weekly*, 40(44–45), 4737–4745.

Moola, S. (2006) 'Women and peace-building: The case of Mabedlane Women', *Agenda*, 20(69), 124–133.

Moser, A. (2007) 'The peace and conflict gender analysis: UNIFEM's research in the Solomon Islands', *Gender and Development*, 15(2), 231–239.

OPAPP (2015) Home page. Available at: http://opapp.gov.ph (accessed 12 November 2015).

Oracion, E.G., Miller, M.L. and Christie, P. (2005) 'Marine protected areas for whom? Fisheries, tourism, and solidarity in a Philippine community', *Ocean and Coastal Management*, 48(3), 393–410.

Pennington, E. (2012) 'Give women a chance to make peace in Mindanao', *Weekly Insight and Analysis in Asia*. The Asia Foundation. Available at: http://asiafoundation.org/2012/04/25/give-women-a-chance-to-make-peace-in-mindanao.

Quindoza-Santiago, L. (2007) *Sexuality and the Filipina*. Quezon City: University of the Philippines Press.

Vellema, S. and Lara Jr, F. (2011) 'The agrarian roots of contemporary violent conflict in Mindanao, southern Philippines', *Journal of Agrarian Change*, 11(3), 298–320.

11

COMMUNITY AND CONFLICT

The case of Rwanda[1]

Marion E. MacLellan

Introduction

The notion of 'community' is a fluid and variable one, whose interpretation is contingent upon a breadth of differing and sometimes contradictory contexts. These can be spatial—from global to regional, from national to more localised 'grassroots' understandings of community—or can derive from cultural and behavioural paradigms. As a consequence, the role played by a 'community' is often contested, and can be left open to critique when expectations are not fulfilled. This is often the case in times of crisis, such as disasters, both natural and anthropogenic, and the impact on, and actions of a 'community' during and after such events is a relevant and important area of study.

Community structure and functions can shape, and can be, in turn, shaped by disasters: reaction, recovery and reconstruction, from structural to the social (including psycho-social). Thus 'community' may reflect an element of 'strength' and 'resilience' implicit in its being, but a critical event can overturn this, shattering and disrupting structures, roles and relationships, sometimes over the long term.

The case of Rwanda presents an interesting and important example of this relationship between community, behaviour and consequence. Civil unrest in the early 1990s culminated in the genocide of 1994, significantly impacting communities, but also 'facilitated' by existing community structures. Consequently, communities have faced many problems, which have challenged accepted definitions of 'community'.

No discussion of Rwanda's story is possible without an acknowledgement of the role of ethnicity which shadows it, a dynamic for conflict and inequality that remains a silent presence to this day, shaping community structures and restructures. Thus the place and role of 'community' in Rwanda are contentious; its capacity to be a catalyst for rebuilding a nation, uncertain.

This chapter will first present a history of the conflict, then an examination of community during and immediately after the conflict. It will finally look at the Rwanda of today, where significant development progress has been achieved.

Rwanda background

To understand the events of the years since 1994, it is useful to consider the geographical and demographic context of the country. Rwanda is a landlocked country of 11,262,564 inhabitants (NISR, 2016), comprising two main ethnic groups—the majority Hutu (approximately 85 per cent) and minority Tutsi (14 per cent). A third ethnic group, the Twa, accounts for less than 1 per cent of the population. The country's population density is 445 inhabitants per square kilometre. Situated in the Great Lakes region of Africa, Rwanda is bordered by the Democratic Republic of the Congo, Uganda, Tanzania and Burundi (see Figure 11.1).

FIGURE 11.1 Map of Rwanda.

Source: Cartography Department, Coventry University, 2016.

Ethnicity

The emergence of ethnic identity was a relatively recent construct forged by changing and contrasting attitudes and policies, initiated by country leaders, and arising from ideology and intents. During the nineteenth century, when today's Rwanda was founded, Tutsis were given leadership positions, establishing them as people of power within the country. Often this identification originated from movement between clans, occupation as cattle owners or herders, or marriage into the other ethnic group. Thus physical or inherited characteristics were not primary determinants of 'ethnic' identity. This was formalised in 1933, in the era of Belgian colonisation, when the ethnicity of Rwandans had to be indicated on identity cards. Ethnic ascriptions were no longer flexible, depending on marriage or the acquisition of cattle. They became permanent, leading to the notion that origin, physical features and class differentiated Hutu from Tutsi, despite shared traditions, faiths and languages. Indigenous governance structures were replaced by those of the Belgians, and Tutsis were rewarded in terms of power and other favours (Prunier, 1995). Thus a level of ethnic antagonism emerged in the country, carved out of inequality and a lack of opportunity for social movement.

As independence movements grew in Africa from the late 1950s, so the political space in Rwanda was forged along ethnic lines, indicative of the main fault lines in the society, economically and socially. The Belgian administration understood that in democratic elections, a Hutu majority population would return a Hutu government, restricting its influence and interests, so changed its allegiance to the Hutu population (Prunier, 1995). In 1959, a Hutu rebellion resulted in the killing of thousands of Tutsis, with little intervention from the Belgians, followed by the first exodus of a significant number of Tutsis to neighbouring countries (200,000 migrating to Uganda alone) (Keane, 1996). Many of these, children at the time, would return in the 1990s to fight as part of the Rwandan Patriotic Front (RPF), the mainly Tutsi force formed of exiles.

After independence in 1962, the Hutu government instituted discriminatory policies against Tutsis, and multiple incidents of violence and killing occurred during the following years. Ethnicity was a means to maintain power, and the government fomented division within the country, blaming continuing poverty on the Tutsis (Survivors Fund, 2008).

Constant economic challenges and the gradual move to openness and accountability within the donor and international aid world were warnings to those leading Rwanda that its administration and structures would have to adapt. In 1990, the RPF invaded Rwanda from Uganda, an event portrayed by the government as an 'ethnic rebellion'; despite a dearth of Tutsis in the country with political or civic power, they were characterised as a group to be feared and hated, sowing the seeds for the future genocide (Keane, 1996, p. 23). The international community brokered unsuccessful ceasefires and agreements, until the strengthening of the RPF forced the then President, Juvenal Habyarimana, to participate in peace talks and to accept the need for multiparty politics. The Arusha Peace Accords of 1993 led to a power-sharing agreement, including a reduction in the powers of the President, among other concessions to the RPF and the Tutsi population.

Genocide

For many Hutus, this was interpreted as a removal of power and privilege, and when Habyarimana was killed in an aeroplane crash on 6 April 1994, on his return from Arusha to Kigali, the belief that Tutsi rebels were to blame inflamed a hostile situation, with the result that an already conceived attack on the Tutsi population was unleashed.

The killings began immediately, facilitated by the administrative structures of Rwandan society; officers at the various levels within Rwandan society—village, cell, sector, district—held lists of key individuals to be targeted. The establishment of road blocks prevented movement of the population; *interahamwe* militia[2] and the President's troops focused on killing those who were considered enemies; the Hutu population was encouraged to participate in the killing of all Tutsis, be they relatives, neighbours, friends or strangers, with radio, in particular, used to incite the murders (Prunier, 1995; Li, 2004). The RPF began moving through the country towards Kigali to assume power and bring an end to the conflict.

A lack of international intervention permitted this genocide of Tutsis and moderate Hutus to take place over a period of 100 days. From 6 April to 16 July 1994, when the RPF finally took control of the country, between 800,000 and 1 million people were killed out of a population of 7 million (UN, 2016b; Survivors Fund, 2008). Some observers have calculated the number to be nearer to 1.925 million (Musoni, 2008). Within the first month, half of the Tutsi population was said to have been killed (Rakita *et al.*, 2003) and 75 per cent by the end of the crisis (Verwimp, 2004, p. 233). During this time, families fled their homes and communities, with over 2 million people making their way to neighbouring countries; thus a substantial proportion of the population was displaced or deceased.

Those not killed were subjected to high levels of violence. While the *interahamwe* initially targeted males, sexual violence was used extensively as a weapon of war: up to 250,000 women were raped, including most of those between 13 and 50 years of age in Kigali, and in some regions all adult and adolescent women who were not killed were raped (UN, 2016a; Layika, 1996). The act of rape in war is usually employed to stigmatise an individual, family or community; however, in this case, objectives also included the desire to impregnate Tutsi women by 'Hutu' fathers, to 'further' the Hutu race, and to infect Tutsi women with the HIV virus.[3] Thus a double tragedy impacted these women: first the rape itself, then the knowledge that they could suffer a slow death from AIDS.

Machete attacks killed and maimed individuals, leaving the latter with long-term disabilities. The violence was perpetrated in the presence of witnesses, often children, which has caused enduring psychological trauma.

Furthermore, this violence was perpetrated by relatives, friends and neighbours—fracturing households, communities and the nation. The task of reconstruction and reconciliation would be challenging and potentially unattainable.

Communities during and post-conflict

At the time of the genocide, Rwanda was divided administratively into 12 pre-fectures, then into sub-prefectures, subdivided into communes, then sectors (USF, 2016). The final level was the cellule, where a community leader (*nyumakumi*) headed ten households. This ordered structure meant that population details were known to the authorities and identities documented, thus facilitating the search for Tutsi victims by *génocidaires*.[4]

These discrete well-organized community units suffered both physical destruction during the genocide and societal fracture, impacting people psychologically and emotionally.

The demographic composition of the country changed substantially due to the genocide—approximately 70 per cent of Rwanda's population was now female, many of whom were widowed (El-Bushra, 2000). Furthermore, the genocide left large numbers of children orphaned or displaced.

Support

Customary expectations of behaviour within the extended family in African society include the care of the orphaned children of siblings, cousins and other family members—the African saying 'there is no word for orphan in Africa' (Roalkvam, 2005; Foster *et al.*, 1997) emphasises this 'shared' raising of children, functioning as a safety net for the community. Yet when high numbers of deaths occur in events which are anthropogenic or natural in origin, this safety net is stretched to its limits. This can occur in conflicts, in natural disasters such as earthquakes, or in infectious disease pandemics, such as the Ebola crisis of 2014–2015; capacity crosses a 'tipping point', and society is incapable of carrying out its expected responsibilities.

HIV/AIDS had caused an increase in single and double orphans[5] in the years preceding the genocide, and this was dramatically augmented after the crisis, with estimates of 1.2 million infected individuals overall, 1 million of whom were orphaned due to the conflict or HIV infection post-conflict (Unicef, 2006).

The sudden loss of considerable numbers of individuals between 15 and 49 years of age—economically active care givers within society—created a vacuum in communities. High numbers of orphans meant a substantial need for carers, which strained traditional community structures. Consequently, children were forced to find other ways of existing—in the many orphanages which emerged, or caring for themselves and other children (usually siblings) in child-headed households, or living on the streets.

Returning to communities where one might see one's attackers each day, was significantly challenging. For all those left vulnerable after the horrors they had experienced, the need for support was essential; these were defined as 'survivors'—people targeted during the genocide because of their ethnic or political situation—and they numbered in the hundreds of thousands.

Expectations of the part played by communities suggest the fulfilment of needs and the offer of support in difficult circumstances. Yet communities are composed of individuals, for whom personal action is not always in harmony with group dynamics or social mores and expectations, in particular in cases such as this. Conflict, both physical and social, fractures societies and weakens social capital, consequently leading to fewer ties and obligations for individuals. This societal dynamic of increased insularity of households within communities meant that the vulnerable were not supported; even when very young children were left alone with no carers, adults were not interested in caring for them in the community (Wolkow and Ferguson, 2001). A traumatised people mistrusted one another, impacting social capital, as confirmed by Rieder and Elbert, who found a relationship between high levels of post-traumatic stress disorder and low levels of social integration within Rwandan communities (2013, pp. 11–12).

Notions of community-based support or care have to acknowledge the fact that communities may be neither benevolent nor cohesive (Thurman *et al.*, 2008): crucial elements in behavioural choice may be inclination and capacity, and so individuals or groups may ignore the needs of others due to ethnicity, historical events (genocidal violence), imprisonment or hierarchy, or owing to a lack of resources—financial and human.

In Rwanda, the vulnerable remained unprotected—females who had survived rape and widowhood, and orphaned children, continued to be subject to attacks and thefts in their own communities instead of being supported and cared for. Relationships had been broken at all levels, thus the expectations of the role of the community were unmet, with the vulnerable left unable to depend on the community for support (Veale, 2000).

Communities were not only riven with social fissures; there were also material and practical challenges to overcome, as all people were enduring various levels of poverty. Consequently, individuals were primarily concerned with the survival of their own households; they were unwilling to relieve, or were incapable of relieving, the suffering of neighbours, whatever their situation.

The dearth of international intervention during the genocide was partially redressed by the arrival of an army of international NGOs and agencies. Although its numbers have diminished over the last 20 years, it remains substantial. NGOs engaged in work in the refugee camps of bordering countries, and proceeded to assist in the rebuilding of basic services in Rwanda itself. Many people benefited from programmes dedicated to housing reconstruction, education provision, healthcare and skills training. These initiatives provided important aid; however, some negative consequences ensued. Resentment grew between those included in the programmes and those who were not—despite experiencing similar or worse hardship. Again, the paradigm of ethnicity was a contributory factor, since the government classification of 'genocide survivor' referred to Tutsis only. Thus, the shadow of ethnic identity shaped even support structures.

Land and property

The RPF's victory heralded the return to Rwanda of refugees not only from the genocide but also from earlier exoduses. This brought with it increasing challenges of land and property ownership, already highlighted as a deep-rooted problem in densely populated Rwanda. The returnees laid claim to land and property they owned according to customary law, initiating large numbers of disputes, land grabs and homelessness for many.

Furthermore, land property law was such that widows and orphans were left dispossessed of land on the death of male relatives, causing family disagreements:

> In our culture we are part of the property. We women are owned by the families. And if you look at our judicial system, 99 per cent of cases are land and property disputes.
>
> (*Aloysie Inyumba, UN-Habitat, 2002, p. 2*)

Extended families seized what was customarily theirs, superseding any instinct of care and support for the vulnerable, no matter what their age.

Thus, if family members or kin were not prepared to support one another, still less likely were neighbours to fulfil this role, with no blood ties drawing them together, and with numerous barriers, including ethnicity and poverty, hampering community cooperation.

Communities were fragmented, and the arrival of new families emphasised and prolonged the lack of trust and community cohesion. The release of prisoners back into the community heralded further waves of mistrust, in particular in 2003, when a significant number were returned to their communities (Rieder and Elbert, 2013, p. 11).

Finally, the lack of housing led to the government initiating 'villagisation' (*imidigudu*) building programmes, whereby families would be rehoused in newly built communities. The primary purpose was to provide homes for genocide survivors and returnees, but it also aimed to improve the system of land distribution and management. *Imidigudu* aimed to encourage the establishment of development centres in rural areas and break with traditional scattered housing. Benefits included improved land utilisation and the provision of basic services (Palmer, 1999). The government saw this as the only long-term solution for Rwanda, since years of conflict by infractions (especially cross-border), riots and looting had destroyed farms and huts. Its goal was for 'all Rwandans to live in *Imidigudu* in urban or rural settings' (IRIN, 2004), where better access to services such as water would be provided.

For some, this would be an opportunity to start their lives afresh, since return to home communities had stimulated memories of violence or renewed feelings of hostility and stigma. Yet others were reluctant to move into the new communities. Criticisms included obligatory villagisation, limited land for crops near to housing, and lack of infrastructure and facilities to enable economic activities (Liversage, 2003). People had to farm on eroding hillsides, and argued that the land policies privileged survivors to the detriment of the landless (Tiemessen, 2005). Former valued social networks, which facilitated income generation and support, such as church groups, were far away.

From the perspective of community acceptance, villagisation might bring advantages—most of the community would be new to an area, so relationships were initiated and communities established from scratch, without the shadow of past experiences. Certainly the experiences of some individuals living in communities purposefully constructed for widows and orphans were positive: one survivor acknowledged that everyone in her community faced the same challenges, so they were able to empathise with one another and offer help and support (MacLellan, 2010). Yet whether such 'artificial' communities facilitate effective reintegration into broader society is questionable—'othering'[6] might reinforce and prolong separateness and dislocation.

The fragility of community cohesion in Rwanda during and post-conflict has many sources—intracommunity violence and relationship breakdown, ethnicity, stigma and poverty. Idealistic expectations of 'community' are challenged, and other contexts reiterate this: research has revealed that child-headed households in Zimbabwe often receive no support from extended family and community, who are ashamed of their existence (Roalkvam, 2005, p. 211).

Recent developments

The post-conflict years have generated further challenges for the concept and function of 'community' in Rwanda. The prospect of closer reconciliation between people, of unity through crisis, was one of the primary narratives of the post-conflict government. To achieve this, President Kagame and his government have promoted a particular political discourse—undertaking a significant task in attempting to erode ethnic differences, alongside a considerable and ambitious programme of economic and social development in the country.

The objective of reconciliation is realised by government policies on ethnicity, with inclusion at the core. The removal of discrimination and attempts to underline unity within the country have been key themes—the identifying factor that 'all are Rwandans' and the prohibition of 'Hutu', 'Tutsi' and 'Twa' as means of identification have been prime directives in this agenda. However, ethnicity silently stalks the policies of the government.

In their quest to achieve these aims, the President and his government stand accused of authoritarianism, and of using the narrative of ethnicity and its context in Rwanda to limit opposition within the country. Waldorf has argued that the imposition of this authoritarian regime acts as a safeguard against genocide, although it effectively replicates the pre-genocide government in its suppression of opposition voices (2007, p. 404).

The notion of 'divisionism' has been at the heart of government ideology and policy, introducing legislation proscribing references to ethnicity and establishing 're-education' camps. These offences are considered as extremely serious and many high-profile individuals have consequently been charged and imprisoned in recent years. Use of the media as a vehicle to incite violence during the genocide has resulted in divisionist laws imposing strict limitations on the freedom of the press; as a consequence many journalists have been jailed, banned or disappeared (Sundaram, 2016).

Furthermore, political space within the country continues to be compromised, with opposition parties struggling to provide critical comment on the current government and its policies without provoking disapproval and censure. Individuals from political and military spheres have been arrested, jailed, disappeared or removed from office (Beswick, 2010). The lack of an effective opposition allows the government to continue its policies unchecked, despite criticism from external bodies and governments, to whom it pays little heed.

The unity paradigm also leads to further challenges for communities. The quest to impose the concept of 'nationhood' on the people by the Rwandan government induces a national silence around ethnicity. Within Rwanda, the community is considered in 'binary terms'; that is, not as a homogeneous space, but as one split into ethnic groups of Hutu and Tutsi. The desire for common ground leaves the fears of division unvoiced, with the risk that these will fester under the superficial veneer of nationhood until a future wave of grievance is unleashed and erupts into another catastrophe (Veale, 2000, p. 238). Legal and societal barriers to discussing ethnicity conceal discourses of identity, history and inequality, which can intensify under the surface.

A further challenge relates to reconciliation, healing and memorialisation. These three are interrelated and interdependent, and each can support or inhibit the others. The process of memorialisation has been employed by the government since the genocide: the establishment of the Kigali Genocide Memorial and others across the country and annual commemorations which take place nationally and internationally all serve to remind people of the past, and to ensure that such events will never occur again (de Bortoli, 2014). They allow people to think of those they have lost, as well as encouraging healing and unity. Yet even this emphasises ethnicity; moreover it can prolong the trauma (Straziuso and Sullivan, 2014). For survivors living in communities alongside the perpetrators of violence, such events can reinforce differences and damage communities further. The need for an acknowledgement of the past for self-healing and renewal brings dissonance to the process of reconciliation and restructuring (Schimmel, 2011). Thus the fault line of ethnicity still permeates communities, albeit under the surface.

Furthermore, the post-genocide years have not been uneventful with regard to international relations. The presence of Rwandan opposition forces in neighbouring countries, *génocidaires* in exile across the world, and the task of rebuilding a nation have weighed heavily on the government and the people. The Rwandan government has faced international condemnation for its role in the conflicts in the Democratic Republic of the Congo, and in 2016 stood accused of training Burundian rebels to overthrow their president (Nichols and Charbonneau, 2016).

However, it would not be possible to discuss post-1994 Rwanda without reference to its progress in the field of social and economic development. While disasters such as genocide inevitably reverse and delay development trajectories, in this case it can be seen as having functioned as a catalyst for a new and revitalised Rwanda. Kagame's leadership, often proclaimed as visionary, has focused on building a new foundation for economic growth with less dependency on external aid and increased foreign investment and production. The Vision 2020 programme incorporates goals of good governance, state efficiency, skilled human capital, a vibrant private sector, world class physical infrastructure and a modern agricultural and livestock sector (MINECOFIN, 2012). Significant progress has been achieved in some areas of development—life expectancy increased from 49.3 years in 1988 (and just 27 years in 1990; World Bank, 2016) to 65.7 years in 2015 (prediction) (NISR, 2016); maternal mortality fell from 1,300 per 100,000 live births in 1990 to 290 in 2015 (WHO, 2016); GNI per capita (PPP) was $1,630 in 2014, from $570 in 1991 (World Bank, 2016); finally, the total number of secondary school enrolments increased from 125,124 in 1999 (World Bank, 2004, p. 203) to 565,312 in 2014 (52.9 per cent female and 47.2 per cent male) (MINEDUC, 2014).

Over the past decade, major construction initiatives in Kigali, comprising the removal of poor-quality housing from certain sectors and the building of new residential areas in different suburbs to rehouse those inhabitants, has been accompanied by fierce criticism from residents and other stakeholders. As with villagisation, post-genocide relocation from established social networks and from places of work, particularly in the informal sector, brings challenges. Significantly, the new housing is more expensive to rent or buy, and is regarded as designed for an emerging elite in the new Rwanda, not for the poor. This ambitious programme has attracted limited international recrimination, however, unlike similar initiatives in other countries. Neither has it induced substantial protest in Rwanda itself, which Goodfellow (2013) interprets as a 'politics of silence', contending that people seek to maintain political stability within the still-fragile nation, to trust in the development agenda, and to avoid conflict and law-breaking.

The demolition of old neighbourhoods has been accompanied by a bold project of reconstruction in the central business district of Kigali, with the aim of building high-rise office blocks and upgrading current buildings. The development of this new capital, built to fit in with the government's vision for the future, has no place for run-down high-density housing, where services are limited and often non-existent. Community cohesion, then, continues to be challenged through government-imposed directives to fulfil its extensive vision for the country.

Conclusion

The interpretation of what 'community' *is* and *does* is a changeable and contested one, and the case of Rwanda has demonstrated the challenges that can befall communities in times of crisis, whether through conflict or natural disaster. The conflicting paradigms of ethnicity, behaviour and expectation, poverty, and psychological and societal trauma conflate and have impeded reconstruction and rehabilitation. Experiences of the genocide and initiatives for the renaissance of a new Rwanda have contributed to a delayed restoration of 'community'.

Actions and reactions on the part of family and community have contested cultural expectations; moreover, intentional actions sometimes impoverish individuals and impair livelihoods without consideration of the consequences.

For many Rwandans, integration and reintegration in communities have been difficult, and hampered by ethnicity and recollection. The role of community and community engagement in countries where conflict has led to a lack of social cohesion is critical (Thurman *et al.*, 2008). The value of strong social networks, the 'social capital' of individuals and communities, must be acknowledged and encouraged. Building such capital as a robust foundation for community relations would enhance reconciliation, reconstruction and reintegration in cases of conflict or natural disasters and is a significant part of peace building. Social capital should be secure and resilient before a crisis, so that it becomes more difficult to break down during and after an event.

In an attempt to avoid the catastrophe of genocide in the future, the government has felt obliged to instigate strict laws to rid the country of the genocidal ideology that stalked it. However, in so doing, it has been accused of restricting freedom, repressing opposition and curtailing open debate, which might have offered a healthier and more reconciliatory path to the rebirth of the nation. One's sense of self (identity) is central, and ethnicity is an intrinsic part of this identity. The challenges of restoring wholeness to the people while acknowledging this, and at the same time avoiding a repetition of the tragedies resulting from the overemphasis of difference in the past, face the government of Rwanda.

Fissures in Rwandan society, wrenched apart by the genocide, led to a breakdown in the structures of communities, both physical and psychological. The nature of the conflict and its consequences left a people traumatised by their experiences and mistrustful of others, resulting in a failure to fulfil the expectations of community, both traditional and instinctive. Allowing a sense of identity, as well as reinforcing an equal and integrated society, would forge a new sense of community within Rwanda, strengthened and resilient in the face of future challenges.

Notes

1 This chapter is informed by research carried out by the author in Rwanda between 2004 and 2010 as part of an unpublished PhD, looking at the livelihood experiences of child-headed households in the country.

2 *Interahamwe*—civilian killing squads from the Hutu population, literally 'those who fight together'.
3 HIV rates rates had reached up to 30 per cent in some areas prior to the genocide (Kayirangwa *et al.*, 2006).
4 *Génocidaires*—those who took part in genocidal killings.
5 Single orphan—death of one parent; double orphan—death of both parents.
6 Othered—from 'othering'—'a process where identities are set up in an unequal relationship' (Crang, 1998, p. 61).

References

Beswick, D. (2010) 'Managing dissent in a post-genocide environment: The challenge of political space', *Rwanda, Development and Change*, 41(2), pp. 225–251.

Bonnier, E., Poulsen, J., Rogall, T. and Stryjan, M. (2015) *Preparing for Genocide: Community Work in Rwanda*. Available at: http://cega.berkeley.edu/assets/miscellaneous_files/83-ABCA-.

Crang, M. (1998) *Cultural Geography*. London: Routledge.

de Bortoli (2014) 'Post-conflict memorialisation in Rwanda and South Africa'. Australian Institute of International Affairs. Available at: www.internationalaffairs.org.au/post-conflict-memorialisation-in-rwanda-and-south-africa (accessed 9 March 2016).

El-Bushra, J. (2000) 'Transforming conflict: Some thoughts on a gendered understanding of conflict processes'. In Jacobs, S., Jacobson, R. and Marchbank, J. (eds) *States of Conflict, Gender Violence and Resistance*. London: Zed Books.

Foster, G., Makufa, C., Drew, R. and Kralovec, E. (1997) 'Factors leading to the establishment of child-headed households: The case of Zimbabwe', *Health Transition Revue*, 7 (Suppl. 2), pp. 155–168.

Goodfellow, T. (2013) 'Kigali 2020: The politics of silence in the city of shock'. Open Security Conflict and Peacebuilding. Available at: www.opendemocracy.net/opensecurity/thomas-goodfellow/kigali-2020-politics-of-silence-in-city-of-shock (accessed 13 March 2016).

IRIN (2004) 'Rwanda: Government implements low-cost housing for returnees'. Reliefweb. Available at: http://reliefweb.int/report/rwanda/rwanda-government-implements-low-cost-housing-returnees.

Kayirangwa, E., Hanson, J., Munyakazi, L. and Kabeja, A. (2006) 'Current trends in Rwanda's HIV/AIDS epidemic', *Sexually Transmitted Infections*, 82 (Suppl. 1), pp. i27–i31.

Keane, F. (1996) *Season of Blood: A Rwandan Journey*. London: Penguin.

Layika, F. (1996) 'War crimes against women in Rwanda'. In Reilly, N. (ed.) *Without Reservation: The Beijing Tribunal on Accountability for Women's Human Rights*. New Brunswick, NJ: Centre for Women's Global Leadership.

Li, D. (2004) 'Echoes of violence: Considerations on radio and genocide in Rwanda', *Journal of Genocide Research*, 6 (1), pp. 9–27.

Liversage, H. (2003) *Overview of Rwanda's Land Policy and Land Law and Key Challenges for Implementation*. Rwanda: DFID/MINITERE.

MacLellan M.E. (2010) 'Child headed households in Rwanda: Challenges of definition and livelihood rights', unpublished PhD thesis, Coventry University. Available at: https://curve.coventry.ac.uk/open/items/f6535af2-e61b-1567-add0-bbdaddcf3169/1.

MINECOFIN (2016) *Rwanda Vision 2020*. Ministry of Finance and Economic Planning. Available at: www.minecofin.gov.rw/fileadmin/templates/documents/NDPR/Vision_2020_.pdf (accessed 13 March 2016).

MINEDUC (2014) *Education Statistical Yearbook*. Government of Rwanda. Available at: www. mineduc.gov.rw/fileadmin/user_upload/pdf_files/2014_Education_Statistical_Year book_.pdf (accessed 13 March 2016).

Musoni, E. (2008) 'Rwanda: Report claims 2 million killed in 1984 Genocide', *The New Times*. Available at: http://allafrica.com/stories/printable/200810040044.html (accessed 10 September 2009).

Nichols, M. and Charbonneau, L. (2016) 'Burundi rebels say trained by Rwandan military'. Reuters. Available at: www.reuters.com/article/us-burundi-rwanda-un-idUSKC N0VD04K (accessed 15 March 2016).

NISR (2016) National Institute of Statistics, Rwanda Government. Available at: www. statistics.gov.rw (accessed 22 February 2016).

Palmer, R. (1999) *Report on the Workshop on Land Use and Villagisation in Rwanda*. Oxford: Oxfam.

Prunier, G. (1995) *The Rwanda Crisis: History of Genocide*. London: C. Hurst and Co.

Rakita, S. *et al.* (2003) *Rwanda, Lasting Wounds: Consequences of Genocide and War on Rwanda's Children*. New York: Human Rights Watch.

Rieder, H. and Elbert, T. (2013) 'Rwanda—lasting imprints of a genocide: Trauma, mental health and psychosocial conditions in survivors, former prisoners and their children', *Conflict and Health*, 7, p. 6.

Roalkvam, S. (2005) 'The children left to stand alone', *African Journal of AIDS Research*, 4(3), pp. 211–218.

Schimmel N. (2011) 'The Agahozo-Shalom youth village: Community development for Rwandan orphans and its impact on orphaned genocide survivors', *Progress in Development Studies*, 11(3), pp. 243–250.

Straziuso, J. and Sullivan, K. (2014) 'Dozens of traumatized mourners carried from stadium as Rwandans mark 20th anniversary of horrific genocide', *National Post*. Available at: http://news.nationalpost.com/news/dozens-of-traumatized-mourners-carried-from-stadium-as-rwandans-mark-20th anniversary-of-horrific-genocide (accessed 10 June 2016).

Sundaram, A. (2016) *Bad News: Last Journalists in a Dictatorship*. New York: Doubleday.

Survivors Fund (2008) *Statistics on Rwanda*. Available at: www.survivors-fund.org.uk/ resources/history/statistics.php (accessed 10 September 2009).

Thurman, T.R., Snider, L.A., Boris, N.W., Kalisa, E., Nyirazinyoye, L. and Brown, L. (2008) 'Barriers to the community support of orphans and vulnerable youth in Rwanda', *Social Science and Medicine*, 66, pp. 1557–1567.

Tiemessen, A. (2005) 'Post-genocide Rwanda and villagisation'. Department of Political Science, University of British Columbia. Annual Convention of the International Studies Association, Honolulu.

UN (2016a) 'Background information on sexual violence used as a tool of war'. Outreach Programme on the Rwanda Genocide and the United Nations. Available at: www.un.org/ en/preventgenocide/rwanda/about/bgsexualviolence.shtml (accessed 16 March 2016).

UN (2016b) *Outreach Programme on the Rwanda Genocide and the United Nations*. Available at: www.un.org/en/preventgenocide/rwanda/education/rwandagenocide.shtml (accessed 9 March 2016).

UN-Habitat (2002) *After the Genocide, Property Rights for Rwanda Women*. Available at: http:// ww2.unhabitat.org/mediacentre/documents/feature9.pdf (accessed 25 May 2006).

Unicef (2006) *State of the World's Children: Excluded and Invisible*. New York: Unicef.

USF (2016) *Gitarama Prefecture, Rwanda*. Available at: http://genocide.lib.usf.edu/rwandan-childrenstestimonies/maps/gitaramaprefecture (accessed 22 February 2016).

Veale, A. (2000) 'Dilemmas of "community" in post-emergency Rwanda', *Community, Work and Family*, 3(3), pp. 233–239.

Verwimp, P. (2004) 'Death and survival during the 1994 genocide in Rwanda', *Population Studies*, 58(2), pp. 233–245.

Waldorf, L. (2007) 'Censorship and propaganda in post-genocide Rwanda'. In Thompson, A. (ed.) *The Media and the Rwanda Genocide*. London: Pluto Press/Fountain Publishers/ IDRC.

WHO (2016) 'Maternal mortality in 1990–2015'. Population Division Maternal Mortality Estimation Inter-Agency Group Rwanda. Available at: www.who.int/gho/maternal_ health/countries/rwa.pdf (accessed 13 March 2016).

Wolkow, K.E. and Ferguson, H.B. (2001) 'Community factors in the development of resiliency: Considerations and future directions', *Community Mental Health Journal*, 37(6), pp. 489–498.

World Bank (2004) 'Education in Rwanda: Rebalancing resources to accelerate post-conflict development and poverty reduction'. Available at: http://siteresources.world bank.org/INTAFRICA/Resources/Rwanda_ED_CSR.pdf.

World Bank (2016) 'Life expectancy at birth, total (years)'. Available at: http://data.world bank.org/indicator/SP.DYN.LE00.IN (accessed 22 February 2016).

12

BENEFICIARY DRIVEN RECOVERY (BDR)

An example from the Solomon Islands

Steve Barton

FIGURE 12.1 Solomon Islands coastal housing after earthquake and tsunami.

Introduction

In 2007, the Western Provinces of the Solomon Islands were subjected to an earthquake measuring 8.1 on the Richter scale. This event triggered a tsunami with waves rising 3–5 metres and reports of 10-metre surges in some areas. At least 36,000 people were affected, with 52 people confirmed dead. About 6,300 houses were damaged or destroyed across more than 300 communities (World Vision, 2013).

The author of this chapter was deployed shortly after this event by the International Red Cross (IFRC) to provide emergency shelter support. Seven months of fieldwork on two large and remote islands resulted in a programme that assessed and quantified the damage and provided a range of shelter material options to affected communities. This approach was an extension of earlier work in Australian Aboriginal communities. In 2008, the author developed for NZAID the shelter recovery programme that is the subject of this chapter, and World Vision, Save the Children and Oxfam implemented this. The NZAID programme design expanded the earlier work and was developed with the express and stated intention of 'restoring community confidence and recovery', rather than providing specific quantities of built houses (NZAID, 2008). The author describes this approach as 'Beneficiary Driven Recovery (BDR)'.

This chapter discusses the method and the issues it seeks to address, and provides working details of how these were applied in the Solomons context. Closing comments describe how this methodology of quantification, resource allocation and transfer to recipient authorship has been applied elsewhere and includes discussion of the community recovery implications.

Section 1: Beneficiary Driven Recovery

Matching resources to needs and capacities

Agencies seek to assist those in less fortunate and more vulnerable circumstances with disaster response, risk reduction strategies, community resilience building, etc., and they are generally committed to identifying needs in the areas where they work. Households which are affected by poverty, misfortune or calamity also actively strive for a safer, better life. However, front-line experience suggests that this confluence of goals is not always working out as well as it might, and programme outcomes may fail to meet ongoing needs. It is clear to the author that beneficiary authored and driven programmes, while challenging, have substantial advantages; however, there is a lack of methodologies to facilitate these processes.

This problem of missed recovery opportunity can be seen as having two major aspects: *What is the nature and extent of a need?* and *How can the available resources* (materials, information, services, etc.) *be equitably and usefully applied?* A third, and perhaps most critical, element is to conduct these activities in a way that *restores and builds the confidence and capacity* of the affected communities, households and individuals.

Phase 1: Getting the scope of activities and the budget right

The nature of impact/damage/need

Under the method outlined in this chapter, the first step is to understand the nature of the impact or damage and categorise this. In the Solomons example, this was achieved by establishing three levels of housing damage. However, this approach could be applied to impacts such as extent of crop or property loss, dislocation, extent of trauma (loss of family member), loss of food stocks, etc. All that is required is a clear definition of the impacts to be measured. It will generally be possible to reach recipient agreement on this at field level.

The extent of impact

Having determined this, the next step is to establish the number of occurrences in the defined categories. How many destroyed houses? How many acres of rice inundated? If, for example, it involves measuring health facilities, this translates as the number of facilities involved showing the type of construction, size and condition of the facility, equipment, stocks, etc.

Application of a proportionate value to each category

The objective of this step is to determine proportionate values between the categories (the financial value is added later). Using the value 1 for the lowest impact, values are established for all the other categories. It may require a known amount of resources to top up a clinic, four times the resources to substantially re-equip and ten times the resources to fully stock. This would give values of 1, 4 and 10.

Temporary immersion by floodwater will disrupt and diminish one harvest season, having minor impact on some farmers and significantly reducing income for others. However, total immersion of land by mud will mean completely new farming procedures, demanding 20 times the resources. This may lead to values of 1, 10 and 30.

Determination the total 'units of impact'

The simple mathematics of multiplying the number of occurrences in each category by its proportionate value leads to the total event impact (expressed as a total number of units). For example:

    600 farmers with a low impact of 1 = 600
    150 farmers with a medium impact of 10 = 1500
    100 farmers with a high impact of 30 = 3000
    Total units of impact = 5100

Applying units of impact to the budget

When the budget is divided by the total number of impact units, the monetary value of one unit can be established. This value is applied to each category: e.g. a value of 5 would be five times the unit value. If this is found to be a suitable figure for each category, all is well. If not, the budget can be modified or the scope changed. This process can also be used from the opposite direction, where the cost of a suitable intervention is extrapolated to arrive at a budget.

After concluding the base work of identifying the nature, extent and value of impact and the shorthand mathematics of steps 4 and 5, the method is easy to apply and remarkably informative. This has distinct advantages, including:

- adapting to expanding and contracting budgets;
- easy demonstration of the budget appropriately addressing (or otherwise) the need;
- being able to see easily how changes to any aspect (budget, categories, affected population) will affect the groundwork;
- easy demonstration of equity and fairness to the receiving group.

Phase 2: Supporting self-determination

Having a clear understanding that that there are different levels of impact, and a variety of resources available, lays the groundwork for self-determination with complete clarity that there is a *range* of needs and range in the *extent of support* available. An example of steps that build on the work of Phase 1 follows.

- *What support will be useful and appropriate?* Investigate what range of materials or services will be useful to the affected population, be they households, individuals or communities.
- *Confirm supply.* Determine which of these resources can be reliably sourced and delivered, checking that this aligns with overarching polices and goals.
- *Produce an available resources list.* List and apply individual costs to all these resources.
- *Confirm the capacity of the recipient to select and apply the resources.* Develop a method to identify the extent (often minor) of technical or other support (including funding) needed by recipients to be able to make best use of the resources. Where additional support is indicated, operationalise its delivery.
- *Offer the resources matched to the category values.* Once the recipients understand the impact categorisation and value that has been established they can select resources up to this value.
- *Delivery and follow up.* As in all programmes monitoring is included with delivery of the resources selected. This should identify households or communities that need further support, seek to capitalise on any opportunities and address unintended consequences.

Reflections

Is all this attention to detail important?

In most emergency response processes, the selection of relief items that are provided to affected people is decided upon by the agency implementing the help. This is appropriate to emergency life saving, but it operates in a very narrow window of time. In the early and longer-term recovery phases, items procured and delivered often fail to strongly correspond to the needs of the affected groups. For example, after the West Sumatra earthquakes in 2009, 52,000 transitional shelters were constructed. However, later programme reviews found that 50 per cent were not occupied (IFRC, 2011). By the time the transitional shelter programmes were underway, households needed repair materials.

Section 2: Beneficiary Driven Recovery in the Solomon Islands

2007 earthquake and tsunami

The NZAID programme referred to above and implemented in the Solomon Islands utilised the BDR method to place recovery decisions in the hands of affected households. The methodology, acknowledging the wide variety of household circumstances, defined damage impact and applied an appropriate value to this. Affected households were provided with a resource list from which they chose (up to an agreed value) materials and services, including a limited amount of cash that most suited their individual recovery plans. A similar process was conducted at community level. The objective was to support households and communities to take the actions that would most support their own recovery, rather than an approach that focused on the output of hardware such as built structures. Every household owning a house damaged by the disaster was able to select materials from a list provided by an aid agency up to a certain budget limit. This limit was set according to the extent of damage that had been done to their shelter: e.g. people whose house had collapsed could claim more materials than those whose house had only been damaged. The aid agency entered into a Memorandum of Understanding (MoU) with each household and community to deliver the materials.

Based on how many of its members were affected by the disaster, and to what extent, the communities were also entitled to a communal budget for construction materials to repair buildings such as schools, churches or clinics. The building process was connected to 'build back better' training activities for the households.

Process

This method made use of the capacities of the individual families and communities to design and build their own shelters and improve community-owned assets. The end product was therefore always culturally appropriate and adapted to the divergent needs of individuals. At the same time, the donors and implementing agencies had quality control over the process by reviewing and approving the list of resources that beneficiaries could choose from. The amount of support was directly linked to the extent of the damage endured by each household. Early in the process, it was clear how much money each household and community would receive, and how much money the donor/implementing agency would spend, thus providing transparency and equity as well as avoiding budget overruns or under-spending.

The project started some time after immediate response activities by several agencies, which had included distribution of shelter items and some limited construction, as well as some training and information activities regarding better building practices and understanding future tsunami risks. At the time of design and implementation, it was not clear what the government response would entail. NZAID accordingly decided to design and implement a shelter/recovery process that would complement any other shelter initiatives. The strategy developed consisted of the following steps.

Establishment of damage categories

During the early response phase, teams from the Red Cross were trained to assess the damaged houses into three categories:

- **Damaged:** The house can be repaired without demolition; it may be leaning, may require re-stumping and will require additional bracing.

FIGURE 12.2 Establishment of damage categories: damaged.

- **Collapsed:** The house has been displaced from its stumps but can be dismantled and the materials re-used. New stumps and bracing will be required.

FIGURE 12.3 Establishment of damage categories: collapsed.

- **Total loss:** The house has completely disappeared or been damaged to the point where the original materials cannot be re-used.

FIGURE 12.4 Establishment of damage categories: total loss.

The categories of damage and which category each dwelling fell into were discussed with individual households and often the wider community at the time of the damage assessment.

There was very little difficulty in reaching common agreement.

Determination of damage impact values

It requires vastly more resources to construct a complete house than to repair a damaged house. To decide a proportionate value for each category of loss in this situation, the volume of timber (the key and most expensive resource needed for recovery) was used as follows:

- A damaged house can be straightened in position. Generally, it will require bracing between the stumps. The volume of timber required for this was calculated at 0.25 cubic metres.
- A collapsed house can be either lifted in position or dismantled and reconstructed. Generally, this will require bracing between stumps, some replacement stumps and possibly sub-floor members. The volume of timber required for this was calculated at 1.25 cubic metres.
- A total loss house must be reconstructed from scratch with all new timbers. The absolute minimum volume of timber required for this was calculated at 6 cubic metres.

Based on these calculations, the relationship between the volumes for each category was established at 1:5:20, meaning it would require five times as many units of materials to retrofit a collapsed house and 20 times as many units of materials to rebuild a total loss house than would be required to repair a damaged house.

Determination of appropriate budget

By multiplying the number of assessed dwellings in each category by the assigned unit value (1, 5 or 20), the total damage value was established. Table 12.1 is an example used for the Western Province (excluding Gizo and Shortlands).

Determination of household support amounts

The amount available for family recovery plans was then divided by total damage units. The unit values were then multiplied using the damage category multiplier

TABLE 12.1 Determination of total units of impact.

Damage category	No. of houses damaged	Damage impact value	Total units
Total loss	234	× 20	= 4680
Collapse	666	× 5	= 3330
Damaged	2748	× 1	= 2748
Grand total	3648		10758

TABLE 12.2 Amount per household.

Loss category	Value	Amount per family recovery plan
Damage	1 × 40	$40
Collapse	5 × 40	$200
Total loss	20 × 40	$800

to set the proportion of the aid budget available for families to use when selecting resources from the Available Resources List. Research had shown that these were reasonable amounts for repair or reconstruction of buildings, i.e. that a house could be replaced with local materials for about US$700.

Establishment of beneficiary lists

The damage assessment carried out by the Red Cross during the immediate response phase had determined the approximate number of affected houses, but not the identities of the owners. Agencies implementing the strategy identified the owners (beneficiaries) in their respective operational areas. Each family was then identified as being part of a community, based on their own identification—usually by geographic location, but sometimes via religious affiliation, tribal boundaries or other culturally appropriate criteria. This was vital for the allocation of resources for Community Recovery Plans.

Making a list of appropriate materials or services

During the beneficiary lists establishment phase, enquiries were also conducted as to what tools, materials or services the families and communities desired. Based on this feedback, the Available Resources List was drafted, defining the range of materials and services the implementing agency would offer to beneficiaries, and applying a cost to each item or service.

The lists in the Solomon Islands included, but were not limited to: tools, equipment, milled timber, sago palm, lashing vine, nails, digging and clearing equipment such as spades, hoes, bush knives, use of a portable mill, use of a chain saw and fuel for mills or chain saws.

These lists were sent to the donors for approval, allowing the donor to fit the activities into larger goals and programmes. For example, for building back better (safer), NZAID required all tools and materials to be in the mid- to high-quality range (e.g. all nails distributed had to be galvanised).

Prior to finalising the Available Resources Lists, agencies had to ensure that all the items would be available from suppliers.

Family Recovery Plans

Families with houses categorised as damaged or collapsed were not expected to provide a specific recovery plan, as in most cases they repaired an existing building, but those who suffered a total loss (or those who had to relocate) were asked to prepare a simple Family Recovery Plan. This document, which was signed by the implementing agency and the beneficiary family, included the resources requested and could be used to monitor the recovery process (e.g. whether the items selected had been delivered). The Family Recovery Plan documents were easy to fill in, but it was recognised in advance that some families or individuals (e.g. elderly or single-headed households) might have difficulty developing a plan. These families received additional one-on-one support.

Another initiative allowed families to 'assign' all or part of their individual benefit to Community Recovery Plans, with some families choosing to allocate resources to the community rather than their personal recovery. Additionally, families had the option of combining their benefit value with those of other families to increase their purchasing power.

Community Recovery Plans

Communities received funds according to the amount of damaged, collapsed and total loss houses belonging to their members. A Community Recovery Plan was required for any community that had suffered ten or more total loss houses. It had almost the same function as the Family Recovery Plan and included a suitable tracking document, which also required submission to the donor for approval.

MoU with community or family

Each family and community then selected the items from the Available Resources List that best met their needs, up to the value of their benefit. The families could choose any quantity of any item from the list, as the list included only items the agency was confident it could source and supply in any quantity requested. Implementing agencies signed an agreement with individual families and communities to supply them with the requested goods.

Delivery of materials (or services) as selected

Items selected were procured, and once delivered to the recipient the MoU was signed off as complete, providing evidence for the donor and agency of a completed activity.

Follow up to confirm progress

Family and Community Recovery Plans were monitored as per the required documentation, and where families or communities were identified as continuing to have difficulties, additional support was provided.

Stakeholders

Community and families

The affected families were the primary stakeholders in this project. Community leaders were considered by all implementing agencies as contributing to the overall success of the programme and were responsible for coordination, assisting families, contributing local knowledge and monitoring programme progress. Local women's associations and faith groups were also instrumental in the practical aspects of community engagement, discussion groups, identification of beneficiaries and coordination with the implementing agencies.

Government

Proposals by implementing agencies to have a cooperative shelter reconstruction plan with the government were not possible due to delays and uncertainty regarding the National Recovery Plan. This uncertainty was a key driver for the design of the Beneficiary Driven Recovery project, which aimed to work in parallel with any government response. Because families and communities were expecting support from the government at a future time, a vital aspect of the plan was its flexibility so that families and communities could blend their own capacity with staged and varied support from both implementing agencies and the government.

Funding sources/donors

The strategy design was funded by NZAID. The implementation was funded by NZAID (with approximately US$1.9 million), AusAID (with approximately US$1.9 million) and Oxfam (with approximately US$620,000). RAMSI (Regional Assistance Mission to the Solomon Islands) provided logistical support.

Implementing agencies

The project was implemented in the Western Province by World Vision (from April 2008 to November 2010), in Choiseul Province by Save the Children (from May 2008 to June 2010), and on Gizo Island by Oxfam (from May 2008 to April 2010).

Other stakeholders in the project included international and local agencies, which engaged in the initial emergency response and contributed to the damage assessment.

Successes

The programme offered the opportunity for affected households to play a more than usually active role in decision making regarding their recovery activities rather than receiving generic packages decided by others. This helped to re-establish a sense of control over what had happened to them. This process was recognised as just as significant as (or perhaps even more significant than) the physical infrastructure. Giving households control over their own recovery process also helped increase their satisfaction with the final result and helped identify existing capacities and resources. While model designs for shelters were available, these were not pressed upon the affected families. For example, some households held off reconstruction while they assembled more materials for their long-term housing goals.

A positive outcome of the project was that some families and communities chose to pool resources, allocating their family benefits to Community Recovery Plans, such as repair of a damaged community church. Some of the communities already had development projects that they were working on and the materials given for recovery were fitted into these plans. The community recovery package allowed the communities to take care of those in the community who were not listed as beneficiaries. To quote the evaluation report of World Vision:

> The project team noticed that the Community Benefit Value mitigated potential tensions between those people in a community who were beneficiaries and those who were not. The project has found that the Community Benefit Value, Relocation Benefit Value and Community Package distribution turn out to be the best solutions for frustrated communities who did not benefit from the project.
>
> (*World Vision, 2007/2008*)

Strong attempts were made to address cross-cutting issues, in particular gender equity, environment and future resilience. For example, the three-person committees who coordinated community recovery plans had to have at least one female member.

Challenges

Some things did not go according to plan. For example, it took a lot of time to verify beneficiary lists. The data assembled during the immediate response needed considerable updating. Working with scattered populations on remote islands created additional challenges. Some of the materials requested were not available from local suppliers and needed to be purchased and shipped from other islands, slowing down the progress of the project and increasing the costs. Other challenges provide lessons for future replication and upscaling.

Donor control versus flexibility

The goal of the project was authorship and ownership by the recipients. While the donor was able to control some factors (such as putting only environmentally sustainable materials on the Available Resources List) other factors were decided by the recipients. For example, the donor could not be sure that houses would be built according to build back better (safer) standards. While training activities were undertaken to increase local construction knowledge, especially with regards to risk reduction, adaptation and increasing individual and community resilience, people were encouraged but not forced to use what they had learned.

Donor unanimity

Donors who joined the programme after it was underway had different objectives than those articulated by NZAID: for example, to promote recovery and restoration of confidence rather than absolute numbers of built houses. As such, it was difficult for these later implementing agencies to report numbers of houses constructed (never a goal of the strategy), as some community members decided to repair their houses incrementally or allocate resources to other aspects of recovery. The author has observed that the process of house construction in non-disaster times can take many years as materials are gradually acquired. This approach challenged the idea of shelter recovery being measured by the number of rapidly built and completed houses. To avoid future friction between project partners using the BDR approach, considerable effort should be made to communicate this departure from a typical shelter programme.

Beneficiary understanding

Many of the beneficiaries expected to receive a complete set of materials and labour that would allow them to build a full house. Like the donors and implementing agencies, they also had to get used to the new way of working and the limitations of the budget. However, the project teams spent significant time building understanding in the communities around the strategy, which contributed to community acceptance and engagement.

Effectiveness of training activities

During the emergency response phase, agencies had conducted well-presented workshops on a variety of design and construction recommendations. However, there were few participants and the recommendations were generally not adopted in the initial recovery phase. This was probably because the workshops attempted to address too many issues. Later in the process, when it had been identified that sub-floor bracing was the primary cause for building failure, training aimed at this specific problem was carried out and many families did start to brace their houses. In the future, when training sessions are given in the early stages of recovery, it may be more effective to focus on one central message.

Section 3: Beneficiary Driven Recovery in other contexts

The BDR framework can be used as a total approach or as elements in a response, as can be seen from the following examples.

Disaster preparedness packaging, Tajikistan, 2008

The international disaster response community has since 2004 developed a 'cluster' system to better coordinate, fund and harmonise its response to major disasters. Under this approach all agencies in each sector (for example, shelter, water and sanitation, nutrition, protection, etc.) meet regularly to achieve these goals. The author was deployed by the IFRC as the Shelter Cluster coordinator within the international response during unusually cold weather.

Natural disaster issues are regular features of life in Tajikistan and long-term agencies have substantial response experiences. To improve on 'one size fits all' disaster response approaches, agencies were asked by the Shelter Cluster to review experience and list items that were successful, not so successful and, in particular, culturally appropriate. Markets were visited to identify, describe and photograph commonly used items. These were listed with values and the plan was for all agencies, such as UNDP, Unicef, Save the Children, ACTED and Oxfam, with long-standing presence in Tajikistan to issue this in booklet form to their outlying offices. In the event of a disaster, affected households could be shown this list and select items to an agreed damage-related value appropriate to their needs.

Recovery from Black Saturday bushfires in Victoria, Australia

In 2009, bushfires devastated Victoria, leading to over 170 deaths and massive property loss. The Victorian Bushfire Recovery Authority received more than AUD$10 million in unsolicited donations. Using the 'Solomon Islands Methodology', a dollar value was applied to each inventory item and proportionate amounts were allocated to beneficiaries to use in selecting items. This helped to resolve the great challenge of allocating and distributing this wide variety of goods and services (VBRRA, 2011).

Tsunami response, Mentawai Islands, Indonesia, 2011

Following a severe tsunami in Western Sumatra and a recovery programming consultancy by the author, Merci Corps reworked its recovery approach and offered affected households the option to select from a range of agricultural tools after it was observed that a standard package was not being used. In the same project, construction of public latrines with volunteer labour was cancelled and households were funded to build individual and culturally appropriate latrines.

Floods response, Myanmar, 2015

In Myanmar, during the 2015 flood emergency the author, again as Shelter Cluster lead, encouraged coordinators in Shelter, Food Security and Livelihoods to form a group approach harmonising cash transfer programmes across sectors and organisations. The guidelines that were developed focused on these three household personal priority sectors, acknowledging that other areas such as water, sanitation, health and education are 'public' activities. The guidelines took into consideration that all households did not experience the same impact from the floods. Some had damaged houses but relatively unaffected agriculture while others had undamaged houses but paddy fields under a metre of silt. The guidelines suggested that the support given to the households should be nuanced to reflect these differing impacts. A matrix was provided that defined a value for the three sectors on the basis of high, medium and low damage. This matrix was also a tool for 'multipurpose' cash for all organisations to adapt and consider households' needs in the three sectors.

Conclusion and relevance to community recovery

In 2007, in the Solomon Islands, towards the end of a long and challenging mission to provide emergency shelter support, the author conducted a comprehensive evaluation of the relief and recovery work he had conducted. On more than one occasion, community members suggested that they might be in a better state seven months after the disaster if aid agencies had never appeared. The provision of support to selected households under a system that was of uncertain 'fairness' and not at all clear to the community members had disrupted the delicate balance of shared assets and responsibilities. Transparency, equity and fairness (as perceived by the affected population) are critical aspects of any effective response.

Communities are usually deeply destabilised by disaster, with day-to-day confidence undermined. The inclination of agencies to assist and advise can deepen a sense of helplessness at the exact moment when confidence in their own capacity needs to be restored.

Beneficiary Driven Recovery seeks to address these issues, first by making it clear that there are varying degrees of loss or impact and by suggesting a method to quantify this. The second step is to make the process of resource allocation both transparent and equitable. Having established this, the process places selection of support squarely in the hands of the communities and their members.

The risks associated with this approach include the very notion of community, the speed of delivery, and transfer of responsibility for recovery to households and communities that may not have the necessary cohesion or technical skills to effect it. The idea that there are homogeneous groups of households living in harmony and accord has been thoroughly challenged by research (as seen in other chapters in this book) and at first hand by the author, including during a two-year period of remote village life. A deep awareness of this truth needs to be brought to bear in any programming.

There is a great temptation and resulting ethos towards urgent delivery of support to people affected by disaster or other calamity. High-speed support is appropriate in some circumstances and this limits the opportunity for the type of collaboration implicit in this methodology. The author's extensive experience suggests that recovery support is usually a longer-term process and the risk of delay is outweighed by the risk of low-value or even inappropriate assistance.

The capacity of beneficiaries, including technical skills, also needs to be carefully considered.

A quotation from Rezki Kandahar, the Shelter Cluster Community Liaison Officer in Western Sumatra, serves as a suitable final comment:

> The mindset of foreign agencies should change and they should realise that they cannot resolve all the housing and other issues of the affected households but only contribute to their bigger aspirations. Discuss people first before shelter and houses. Ensure that what is delivered does not waste time, energy and resources of the agencies themselves—or of the beneficiaries. Do not view the affected families and communities as ignorant victims only, as incidentals needed to run programmes and to get funding. Do not burden the already burdened with something they do not need or want.
>
> (IFRC Indonesia Earthquakes Shelter Cluster Review Jan 2011)

References

IFRC (2011) *Indonesia Earthquakes Shelter Cluster Review*. Geneva: IFRC (International Federation of Red Cross and Red Crescent Societies).

NZAID (2008) *Shelter Recovery Assistance Strategy*. Honiara: NZAID.

VBRRA (Victorian Bushfire Reconstruction and Recovery Authority) (2011) *Legacy Report*. Available at: http://trove.nla.gov.au/work/159343996?selectedversion=NBD51051068 (accessed 31 March 2014).

World Vision (2007/2008) *Solomon Islands Project Documents*. Burwood, Victoria: World Vision.

World Vision (2013) *Humanitarian Emergency Assistance Factsheet: Solomon Islands*. Burwood, Victoria: World Vision.

13

FROM SHORT-TERM RELIEF TO THE REVIVAL OF COMMUNITY IN POST-TSUNAMI SRI LANKA AND INDIA

Martin Mulligan

AUTHOR'S STORY

In order to conduct a study which could unearth some examples of good practice in rebuilding local communities after the December 2004 tsunami disaster in Sri Lanka and southern India, the author travelled extensively before selecting study sites, often with co-researcher Yaso Nadarajah, and consulted many community-based organisations. In the Ampara District of eastern Sri Lanka, we relied heavily on the local knowledge of regional community development worker Ashraff Mohammed and, following a long journey down the coast, we arrived in a resettlement village that Ashraff had heard about, called Kudilnilam. Ashraff was keen to see what was happening in this rather remote settlement because he had heard that the lead agency, the People's Church of Sri Lanka, was doing a good job. Having seen debates in the Sri Lankan media about the danger of religious organisations using tsunami aid to win converts for their religions, we were rather apprehensive about what we would find. It was immediately obvious that the completed new houses were solidly built and were laid out as a village surrounding a common 'green space' and community centre. At the centre we were met by 'Pastor Ram', who told us that when news of the devastation caused by the tsunami reached the People's Church congregation in Colombo, they decided to load up a truck with food and equipment and head for war-affected Ampara, because it was clear that this district would not get as much aid as the southern province. They kept driving until they reached Thirukkovil, where little aid had arrived, and began working with families taking shelter in a local school. With no prior experience in this kind of work, Pastor Ram and his volunteers patiently tried to respond to needs as they saw them, and they were asked by local authorities if they would take responsibility for overseeing the construction of a resettlement village. They insisted on knowing which families would be allocated housing in the new village so that they could work with them on village design and house construction. When we first visited Kudilnilam, Pastor Ram and his volunteers had already worked with many of the families there for two years and when we returned several more times over the next two years, it became apparent that the Colombo-based church volunteers would stay as long as it took to ensure that the village had been completed to the satisfaction of the residents and that all required services—such as a bus connection to the main town—had been established. The People's Church volunteers were highly regarded by residents we interviewed and they insisted that absolutely no pressure was put on any of them to join the church. Indeed, the volunteers had helped the residents to establish a small, if rather makeshift, Hindu temple in the village. As this chapter shows, we were able to pick Kudilnilam as an example of good practice, even though the aid agency had no prior experience in disaster relief and recovery work. They did have enormous patience and empathy, and a willingness to listen. They worked with individuals who had a capacity to become community leaders and progressively handed responsibility for the functioning of the village to a village committee. This story suggests that community development disaster recovery work is not rocket science, but it does require patience and empathy. It suggests that aptitude for the work may be more important than previously acquired skills.

Introduction

The Indian Ocean tsunami of December 2004 shocked the world because it had devastating consequences for so many local communities, spread across four different nations. The global aid response was unprecedented in terms of the volume of aid given and the number of people and organisations involved in disaster recovery work. For Sri Lanka alone, it has been estimated that more than 500 international aid agencies were involved in post-tsunami relief and recovery work (Silva, 2009). Some reviews suggested that while quite a lot was learnt from the experience about how to mount quick and effective relief operations, many mistakes were made in working with local communities (see, for example, Telford and Cosgrave, 2007; Barenstein and Iyengar, 2010). Indeed, there is little evidence to suggest that much was learnt from the post-tsunami experience about effective strategies for long-term social recovery. This chapter will argue that community engagement in disaster relief should focus, primarily, on transparent assessment of, and response to, complex and sometimes competing needs. Speed and efficiency are critical in the immediate aftermath of a major disaster. However, more patient and inclusive forms of consultation and engagement are needed when the focus starts to shift to long-term recovery because mistakes can have long-term consequences for people and entire communities. Unfortunately, the research on which this chapter is based suggested that many humanitarian agencies failed to appreciate the difference between relief and recovery and many withdrew without putting any kind of 'transitional arrangements' in place (Mulligan and Nadarajah, 2012).

In presenting a summary of the review of post-tsunami relief and recovery work by the London-based Tsunami Evaluation Centre, Cosgrave (2007) argued that not enough had been done to put affected communities in the 'driving seat' of relief and recovery operations. Others have suggested that a stronger community development approach would have produced much better outcomes (e.g. Hettige, 2007; Kenny, 2007). However, such proclamations have often been made on the basis of rather thin or narrow evaluations of the post-disaster experience. It is one thing to find some examples of more 'participatory' aid projects and programmes and another to examine the complex challenges facing local communities in more depth and across a wider range of local experiences. Of course, the term 'community development' itself has been contested by some scholars, who argue that it has essentially become an instrument of state control. Furthermore, it is clearly very difficult to work with communities that are traumatised after severe losses. However, it is not difficult to distinguish between 'top-down' and 'bottom-up' community development practices, and the study reported in this chapter set out to find and analyse examples of 'good practice'.

A problem for selecting 'good practice' is that there is little agreement in the disaster management field about what might constitute a good outcome for traumatised local communities. Acting as a UN Special Envoy for Tsunami Recovery, former US President Bill Clinton popularised the suggestion that generous aid contributions meant that it would be possible to 'build back better', and in 2009 he released a report claiming that this had been achieved. However, as Khasalamwa (2009) has argued, prevailing 'asset replacement' strategies in post-disaster aid work can never result in 'build back better' in a social recovery sense. It is not hard to quantify 'asset replacement' outcomes but much more difficult to 'measure' social recovery. The study discussed in this chapter suggested that pre-disaster vulnerabilities of affected communities can be addressed during recovery work and that a stronger sense of community sometimes results from the response to the crisis. In this sense, the study found that 'build back better' is possible but only rarely achieved. The study suggested that an effective community strengthening strategy requires skill and particular aptitudes on the part of 'external' relief and aid workers. Above all, however, it requires patience and acute sensitivity to local circumstances. In arguing that there are very distinct stages in a successful transition from short-term relief to long-term recovery, this chapter is arguing for a much more 'deliberative' approach to disaster recovery work.

A range of important studies has been published since the completion of the study discussed in this chapter. For example, the renowned Sri Lanka cultural historian Dennis McGilvray collaborated with the anthropologist Michele Gamburd (2010) to publish a collection of essays on the post-tsunami recovery effort in Sri Lanka in order to make the point that there needs to be much more emphasis on understanding local cultural contexts in post-disaster recovery work. Lyons *et al.* (2010) also pulled together an important collection of essays—by both academics and practitioners—to develop the arguments for 'people-centred' or 'owner-driven' housing reconstruction in post-disaster situations. The latter book is not confined to an analysis of the post-tsunami recovery, but it does include reflections on that experience in order to drill down into the rhetoric about 'community participation' in disaster recovery. Lyons *et al.* make it clear that many problems could be avoided if aid agencies took the time to consult and engage affected communities more thoroughly, but they do not drill down into the complexities of 'participation' and 'engagement' in the way that the study reported in this chapter did. While most of the post-tsunami studies have focused on one particular aspect of recovery work—such as housing or livelihoods—this study took a broader approach and the findings confirm the need for better integration of physical and social planning in recovery operations (Mulligan and Nadarajah, 2012).

The study

The study reported in this chapter was conducted over a period of four years in five case study areas (see Table 13.1).

TABLE 13.1 Key characteristics of study sites.

Study site	Key characteristics
Seenigama	A cluster of villages located near the regional centre of Galle and on the outskirts of the popular tourist town of Hikkaduwa in south-west Sri Lanka. Located near an historic Buddhist temple. Almost entirely Sinhalese Buddhist.
Urban Hambantota	An old settlement at the heart of the Hambantota District in south-east Sri Lanka, renowned for its fishing harbour, a major salt industry, nearby national parks and a nearby temple that attracts both Buddhist and Hindu pilgrims. Located on the ancient 'Silk Road of the Sea', at the time of the tsunami the population was nearly half Tamil-speaking Muslim and half Sinhalese Buddhist. Very severely impacted by the tsunami.
Thirukkovil	A large settlement in the Ampara District of eastern Sri Lanka, clustered around an ancient Hindu temple. Badly affected by the civil war. Almost all of the population is Tamil-speaking Hindu. The study focused on relocated communities, settled some 3–4 kilometres from the old town.
Sainthamuruthu	A densely packed settlement in the Ampara District of eastern Sri Lanka, built on a fairly narrow strip of coastal land and adjacent to an ancient and important Muslim mosque. Community almost entirely Tamil-speaking Muslim. Very severely impacted by the tsunami.
Northern Chennai	People from eight devastated coastal 'shanty towns' were first relocated into a big, single temporary settlement and then into two adjacent permanent settlements, several kilometres from the sea. Predominantly low-caste Hindu people who had depended on fishing for their livelihoods.

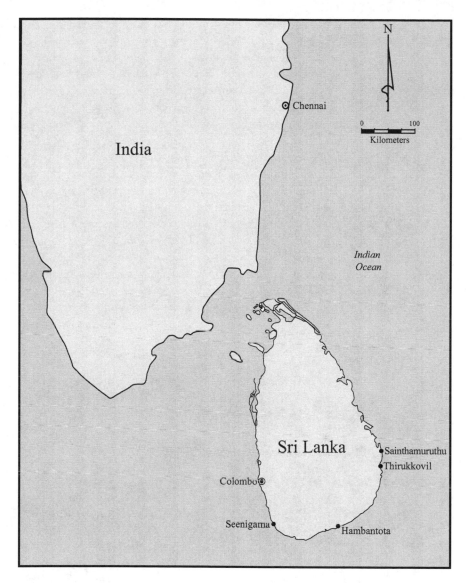

FIGURE 13.1 Map showing location of study sites in Sri Lanka and southern India.

The study was conducted by the RMIT University researchers Martin Mulligan and Yaso Nadarajah, with assistance from Sri Lankan research associates and ten fieldworkers from the communities concerned. The researchers worked in consultation with a number of community-based organisations, such as the Foundation of Goodness (Seenigama), the Hambantota District Chamber of Commerce, and the Natural Environment and Sustainable Development Organization (based in Sainthamuruthu). Surveys and interviews were conducted in a mixture of English,

Sinhalese and Tamil. The study included a random 'community life' survey, the collection of 'community member profiles' (which focused on post-tsunami experiences), the collection of relevant local stories, and a series of lengthy semi-structured interviews with tsunami survivors and a wide range of people working in relief and recovery projects and programmes. Demographic data, accounts of local history, and other reports and stories were collected in order to construct a 'social profile' of each case study community to make sure that the researchers could properly contextualise their findings.

The study was able to analyse the good work of the Foundation of Goodness (FoG) at Seenigama in southern Sri Lanka; the work undertaken by the Taiwan-based Tzu Chi Foundation at Hambantota, also in southern Sri Lanka; the development of a 'model village' at Hambantota by an organisation called Sri Lanka Solidarity which was set up by expatriate Frenchman Philippe Fabry; and the community development efforts undertaken by the Colombo-based People's Church at a remote community near Thirukkovil in eastern Sri Lanka. In each of these cases, the lead agency made sure that housing reconstruction was embedded within broader community development plans, and each of the lead agencies adopted a patient and careful approach in ensuring that design and construction of

FIGURE 13.2 Seenigama Perehera 2007: the Foundation of Goodness recognised that the revival of an important religious and cultural festival would boost community morale and resilience.

resettlement villages could meet diverse community needs. The study also came across examples of effective local action by resilient local communities—as, for example, when tsunami survivors, relocated to Thillagar Nagar in Chennai, India, managed to build their own temple with very few resources. FoG used tsunami aid to build on earlier initiatives to set up community centres, services and facilities which reflected a deep knowledge of community needs and aspirations. FoG also saw the need to rebuild social and cultural activities as well as infrastructure; a good example of this was the revival in 2007 of an important religious and cultural festival which really boosted overall community morale (see Figure 13.2).

Sadly, examples of good practice were hard to find because most resettlement was undertaken on the basis of very poor physical, economic and social planning, and when most international aid agencies wound up their operations after two years few arrangements were in place for ongoing community development work or for local and regional economic development to build more secure livelihoods. Most of the vulnerable coastal communities that were worst affected by the tsunami were left with similar vulnerabilities after the aid effort ended. Indeed, the researchers found that many people were left in even worse circumstances, most notably in the Muslim community of Sainthamuruthu in eastern Sri Lanka, where more than 400 families were still living in squalid 'temporary shelters' five years after the disaster, with no serious prospect of a decent house.

Rebuilding for long-term needs

The study found that in southern Sri Lanka there was considerable wastage and duplication in regard to aid delivery, because of inadequate coordination among the many international and national aid agencies that were involved in relief and recovery work, and because there was little transparency in regard to the assessment of needs and the compilation of lists of people to receive aid (beneficiaries). There was evidence of competition between aid agencies and in some cases, notably in Hambantota, unnecessary haste in constructing new permanent houses resulted in poor planning and shabby construction. Across our case study areas, it was only in the Tzu Chi Great Love Village in Hambantota—built under the supervision of the Taiwan-based Tzu Chi Foundation—that attention was paid to the construction of kitchens that would suit the needs of local families.[1] The research found that women and children often faced particular difficulties when living in hastily constructed temporary shelters, and women with small children were often desperate to move into permanent new housing. However, poorly planned housing allocations often meant that such women were placed in rather isolated settlements with few communal facilities and little in the way of social support. For example, widows with young children were put into a village in the 'new town' of Hambantota that was in the path of travelling wild elephants, because they were put at the top of a priority list for relocation by district authorities. Relocation often made things more difficult for the disaster victims and more effort should have been made to rebuild housing close to original settlements, the research found.

FIGURE 13.3 Haste and a failure of social planning led to the construction of unusable houses in the 'new town' of Hambantota.
Source: Author's photo.

A key recommendation emerging from this research is that there needs to be a more 'deliberative' process in shifting the emphasis from short-term relief to longer-term resettlement. Obviously, relief work needs to be carried out with great speed and efficiency. However, much more thought and planning should have gone into the construction of 'temporary shelters' in which people lived for up to four years after the disaster. If people had been placed in more adequate temporary shelters, then more care could have been taken in the physical and social planning of the new permanent settlements. A great deal of thought went into the planning of Istouti Village at Hambantota, constructed under the supervision of the newly formed NGO Sri Lanka Solidarity. As SLS founder Philippe Fabry told the researchers, recovery workers should be imagining how the multifaceted community might be operating in 20 years' time rather than just putting people into houses built in unimaginative rows in a paddock. However, the extra time taken to plan Istouti Village meant that it took longer to complete,[2] and tsunami survivors who were deemed to be the 'most needy' by district authorities were commonly relocated into inferior settlements before better ones were ready. As a result, tsunami survivors have ended up with housing assets of very different quality and value, through no fault of their own. Poorly constructed houses will require more maintenance, but often international aid agencies did not make transitional arrangements to ensure that local, regional or national authorities would assume responsibility for maintaining new facilities and infrastructure. A problem in Sri Lanka was that there were very few qualified tradespeople available for the reconstruction work. However, the Tzu Chi Foundation offset this problem by bringing in an experienced construction supervisor and the People's Church gave agreed building specifications to those who would occupy the houses so that they could monitor the work on a daily basis.

Local knowledge

Some international aid agencies established effective partnerships with local people and local community-based organisations. For example, the Irish NGO GOAL set up its headquarters in badly damaged Sainthamuruthu and employed local staff, while other agencies tried to operate from far away Ampara township and relied on imported staff and volunteers. However, the external agencies were often constrained in what they could do to address local needs. For example, the declaration

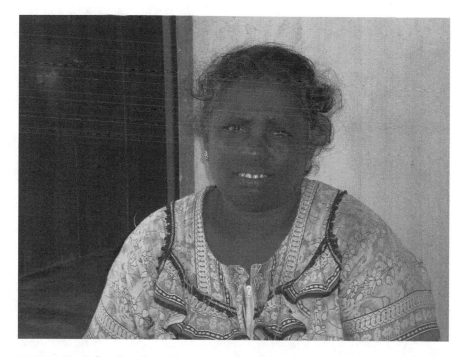

FIGURE 13.4 Cultural neglect: This woman found that in the new relocated settlement she would not be able to provide a house for her daughter on marriage, as expected in her culture.

Source: Author's photo.

by the national government of a 200-metre 'no-build buffer zone' for eastern Sri Lanka meant that it was extremely hard to find land to resettle tsunami victims in Sainthamuruthu (even though the buffer was subsequently reduced to 65 metres in this area), so GOAL had to focus its work on public buildings and infrastructure. Our research confirmed the findings of a study by Thurnheer (2009) in regard to the inappropriate allocation to men only of title to new permanent houses, when traditional patterns of matrilineal descent mean that families are obliged to provide houses for their daughters in order to secure appropriate marriages. Women who had been relocated to Mandanai village near Thirukkovil told the researchers that they had lost hope of securing good marriages for their daughters because they would not be able to provide houses for them in the village where no new building would be allowed (see Figure 13.4).

Of course, some international aid agencies—and it is estimated that around 500 were involved in post-tsunami relief and recovery in Sri Lanka alone (Silva, 2009)—did little to seek out local knowledge. Leaders of the small, local, Al Hikma Foundation in Hambantota said they had to work day and night for over a year to sort out the allocation of donated goods that arrived in truckloads to their town, without any distribution arrangements in place, in order to prevent competition and division within the community over inequitable distribution of such aid. Rather ironically, Muslim Al Hikma worked hard to secure aid funding for the reconstruction of a badly damaged Buddhist temple that was otherwise being neglected. Al Hikma representatives served on a civil society committee established by the Hambantota District Chamber of Commerce which worked diligently for 18 months to ensure that relief and recovery aid was distributed fairly to all affected people. Over half the people who lost their homes were from the local Muslim community, yet the civil society committee was determined to ensure that recovery work would not favour the Sinhalese people in a district in which they formed a big majority. This effort was rewarded when Sinhalese residents helped to prevent the planned relocation of an historic mosque which was badly damaged by the tsunami.

Patience and transition arrangements

When the researchers first began fieldwork for this study in southern Sri Lanka about 11 months after the disaster, they were told by a number of people working for both international and Sri Lankan NGOs that the time had come to shift from relief to community development. A year later, many of the INGOs were closing down their relief and recovery operations because they had begun with the assumption that they must not go beyond two years in order to avoid any growth of 'aid dependency' in the recovering communities. However, such rigid and predetermined timeframes rarely reflected the conditions on the ground and many INGOs left without adequate transition arrangements in place. A number of national Sri Lankan NGOs adopted similarly inflexible timeframes and processes and it became apparent to the researchers that in seeking to become 'professional' aid agencies they had adopted the processes and 'culture' that were prevalent among INGOs. The Taiwan-based Tzu Chi Foundation adopted a longer-term perspective for its work with tsunami survivors in Hambantota. They established a national office for the foundation in Colombo and appointed a local man to be an ongoing coordinator for their work in Hambantota. That man was still in this job five years after the disaster.

A more unexpected example of long-term commitment came from the Colombo-based People's Church, who began their work in and around Thirukkovil by loading a truckload of aid donations as soon as they heard of the disaster and heading off to find a needy community. Pastor Ram Arumugam and his team of volunteers drove along the shattered coast until they found a place where relief was in short supply and began to care for people taking shelter in a school at Thirukkovil. Eventually the People's Church let the district authorities in Thirukkovil know that they wanted to take responsibility for building a new settlement for many of those who could not return to their previous houses, and soon they were given a list of 75 families to house. Pastor Ram and his team began doing community development work with the nominated families while they were still living in temporary shelter, and as soon as possible they told the families which house they had been allocated in the new village so that they could help to ensure that agreed construction standards were met by the builders. After the first 75 families were settled, Pastor Ram and his team began working with another group of 75 families; and when the second set of houses was ready for occupation, the first group of settlers helped the new arrivals integrate into the established community. The Kudilnilam village included a community centre with meeting rooms and a classroom for lessons in computer literacy, and public space for people to meet and for children to play sport. Five years after the tsunami, many responsibilities for dealing with the district authorities had been handed over to elected representatives within the Kudilnilam community, but the People's Church volunteers said they would stay with the community as long as they were convinced that they could be helpful. In general, community morale in Kudilnilam was much higher than in nearby Mandanai, and it was clear that this was largely because the People's Church had made a long-term commitment to community development work.

Of course, INGOs that are trying to respond to disasters all over the world cannot necessarily make a long-term commitment to any particular local community. However, they should try to find other—perhaps local—organisations that can take over such a responsibility, and they need to make sure that adequate transitional arrangements are in place before they leave. Community development needs to begin as early as possible, once traumatised disaster victims feel reasonably safe and settled, and it cannot proceed according to a fixed timetable. Without an adequate commitment to community development, pre-existing divisions and tensions are likely to get worse in the aftermath of a disaster. This was clearly the case in Sainthamuruthu, where young Muslim men, frustrated by the neglect of their community, became a target audience for Muslim extremists. Natural disasters can aggravate social divisions and conflicts, and with climate scientists predicting an increase in their frequency and intensity, all nations should be concerned about the long-term social disturbances that they can trigger.

Focusing on the needs of the most vulnerable

The study found that inadequate attention was paid to the particular needs of women, children and young people. This neglect started in the temporary settlements, which were commonly poorly equipped to meet the needs of women and children, but it continued through to the allocation of new houses and a lack of provision of community facilities, such as play areas for the children. An emphasis on restoring past livelihoods did not cater for the employment needs of young people entering the labour force. While some aid agencies offered 'psycho-social' counselling and some encouraged young children to share their experiences through art or storytelling, there is little evidence to suggest that such projects and programmes were based on any real expertise in regard to the ways in which children react to serious trauma. Some parents told the researchers that their children had begun to display wild and erratic behaviour a long time after the initial trauma, and they suspected that such behaviour could have been linked to the traumatic experience. Clearly, there are experts in many parts of the world who have studied the long-term effects of severe trauma on children, and more should have been done to enable such experts to educate aid workers on how to recognise and deal with indirect manifestations of trauma. The current research did not focus on the way that children reacted to the tsunami trauma, but it suggests that Balaban (2006) is right in saying that this is a neglected field of inquiry in regard to the human impact of disasters in general.

The study found that much of the aid given to help vulnerable people and households re-establish household incomes was poorly targeted. In Seenigama, for example, households were given equipment—such as three-wheelers, sewing machines and refrigerators—to begin new micro-enterprises in the village. However, most of these micro-enterprises failed because there was not enough money in a community that had previously relied on illegal coral 'mining' for livelihoods to support a proliferation of local businesses. Too many small grocery shops were set up, for example, for them all to survive. The Foundation of Goodness (FoG) in Seenigama and the Hambantota District Chamber of Commerce (HDCC) were able to demonstrate that it was better to establish slightly larger enterprises that could clearly fill an identifiable 'niche market', and such enterprises might be able to provide employment for a number of community members. For example, FoG set up a diving school so that former coral miners could obtain qualifications for underwater work in Sri Lanka and the Middle East, and HDCC supported local businesses that employed five to 15 people.

Much more global attention focused on post-tsunami recovery in Sri Lanka than in India, even though the disaster claimed more than 12,000 lives and left more than 130,000 people homeless in India. In part, this was because the government of India announced that it had plenty of experience in disaster recovery and would need little international help. Much of the reconstruction effort was left to the state government of Tamil Nadu and it adopted a top-down approach to the design and construction of resettlement villages (Barenstein and Iyengar, 2010). In northern Chennai, the Tamil Nadu government saw the tsunami outcome as an opportunity to implement a 'slum clearance' plan, and it charged the Tamil Nadu Slum Clearance Board with the responsibility to relocate those who had lost their homes in coastal 'squatter settlements' into medium-rise housing blocks several kilometres from the sea. This effectively ruled out any meaningful community consultation or participation, which, in turn, made community development outcomes almost impossible. It is interesting to note that a range of Indian NGOs used international aid funding to set up Self-Help Groups (SHGs) for tsunami survivors while they were living in a very large temporary settlement in northern Chennai, awaiting the construction of their permanent new homes. Such SHGs were formed on the basis of low-interest loans that could be used for income-generating activities and our study found that they had some beneficial effects. However, this promising initiative was not followed up when the disaster survivors were relocated into the permanent housing blocks, by which time support for generating household incomes had become even more important.

Conclusions

The tsunami waves that tore through communities in so many local areas in four different nations on 26 December 2004 sent shockwaves of concern around the world and these, in turn, generated an unprecedented wave of compassion, resulting in record levels of aid funding, channelled through national governments and international NGOs. Because aid funding flooded into nations like Sri Lanka and India, which did not even realise they were vulnerable to such a disaster and had no significant plans in place for dealing with it, there is no doubt that much of the aid money was wasted. However, rather than focus on what went wrong, the study reported in this chapter sought to find examples of good practice that might inform the work of international relief and aid agencies. In a sense, post-tsunami recovery work was like a big experiment and it is important to learn the lessons for a world anticipating an increase in the frequency and intensity of natural disasters.

In summary, the immediate post-tsunami relief operation went better than expected in countries—such as Sri Lanka and India—which had not anticipated this kind of disaster. However, it was difficult to find examples of good practice in making the transition from short-term relief to long-term social recovery. Of course, as Ingram *et al.* (2006) have noted, it is not easy to balance short-term and long-term needs in disaster recovery work. Yet the study reported in this chapter suggests the need for a process that makes a clear shift from the early emphasis on relief and safety to longer-term planning for resettlement. This, in turn, suggests the need for a more 'deliberative' strategy that is not constrained by predetermined and inflexible timeframes, and which envisages clear and distinct phases in the move from short-term relief to long-term recovery. The long-term aim should be to make sure that people are settled in secure and supportive physical and social environments and to also address some pre-existing vulnerabilities and divisions within affected communities. All local communities contain a host of sub-communities and global flows of people, ideas and influences mean that they are more fluid than ever before. It is important to think of communities as processes rather than entities.

Local knowledge is critical, and it is important that external aid agencies find ways to work with local people and community-based organisations in regard to resettlement planning. Effective partnerships can be forged between external agencies and community-based organisations, and all aid agencies should make sure that they have put in place adequate transitional arrangements before they withdraw from the communities in which they have worked. In short, the study recommends a more deliberative approach in moving from short-term relief to long-term recovery.

Acknowledgements

This chapter is based on the outcomes of a study funded by an Australian Research Council Linkage grant for which AusAID was the 'linkage' partner.

Notes

1 The Tzu Chi Foundation built a total of around 800 houses out of 2,330 in Hambantota's 'new town'.
2 As a condition of being given land for Istouti Village, SLS had to have some houses ready for occupation within two years of the disaster, but construction continued for a further two years after that.

References

Balaban, Victor (2006) 'Psychological assessment of children in disasters and emergencies', *Disasters*, 30(2), pp. 178–198.
Barenstein, Jennifer Duyne and Iyengar, Sushma (2010) 'India: From a culture of housing to a philosophy of reconstruction', in Lyons, Michael, Schilderman, Theo and Boano, Camillo (eds) *Building Back Better: Delivering People-Centred Housing Reconstruction at Scale*. London: Practical Action Publishing.
Cosgrave, John (2007) *Synthesis Report: Expanded Summary, Joint Evaluation of the International Response to the Indian Ocean Tsunami*. London: Tsunami Evaluation Centre.
Hettige, Siri (2007) *Tsunami Recovery in Sri Lanka: Retrospect and Prospect*. Colombo: Social Policy Analysis and Research Centre, University of Colombo.
Ingram, Jane, Franco, Guillermo, del Rio, Cristina Rambaitis and Khazai, Bijan (2006) 'Post-disaster recovery dilemmas: Challenges in balancing short-term and long-term needs for vulnerability reduction', *Environmental Science and Policy*, 9, pp. 607–613.
Kenny, Sue (2007) 'Reconstruction in Ache: Building whose capacity?' *Community Development Journal*, 42(2), pp. 206–221.

Khasalamwa, Sarah (2009) 'Is "build back better" a response to vulnerability? Analysis of the post-tsunami humanitarian interventions in Sri Lanka', *Norwegian Journal of Geography*, 63, pp. 73–88.

Lyons, Michael, Schilderman, Theo and Boano, Camillo (eds) (2010) *Building Back Better: Delivering People-Centred Housing Reconstruction at Scale*. London: Practical Action Publishing.

McGilvray, Dennis B. and Gamburd, Michele R. (eds) (2010) *Tsunami Recovery in Sri Lanka: Ethnic and Regional Dimensions*. New York: Routledge.

Mulligan, Martin and Nadarajah, Yaso (2012) *Rebuilding Communities in the Wake of Disaster: Social Recovery in Sri Lanka and India*. New Delhi: Routledge.

Silva, Kalinga Tudor (2009) '"Tsunami third wave" and the politics of disaster management in Sri Lanka', *Norwegian Journal of Geography*, 63, pp. 61–72.

Telford, John and Cosgrave, John (2007) 'The international humanitarian system and the 2004 Indian Ocean earthquake and tsunami', *Disasters*, 31(1), pp. 1–28.

Thurnheer, Katharina (2009) 'A house for a daughter? Contraints and opportunities in post-tsunami Sri Lanka', *Contemporary South Asia*, 17(1), pp. 79–91.

14

COMMUNITY RECOVERY AND THE ROLE OF EMERGENT ORGANISATIONS IN POST-DISASTER HOME BUYOUTS

A case study of Oakwood Beach, New York, United States of America

Alex Greer and Sherri Brokopp Binder

Introduction

Disasters challenge the efficacy of established protective measures society trusts, exposing failures to adapt to surrounding natural and social environments (Oliver-Smith, 1996). When disasters affect communities and reveal these deficiencies, stakeholders (including government, businesses and citizens) must decide how to react and ultimately address unacceptable risk. In the literature, we see three primary methods for managing hazards: structural mitigation to divert the hazard, redistribution of the financial burden of repeat losses through insurance, and relocation of at-risk properties outside of hazardous zones (Perry and Mushkatel, 1984, p. 155). Relocating properties, while a seemingly logical solution, is not easy. Not only is it costly at the front end, the window for meaningful change is narrow. Support for such measures is often strongest immediately after an event, and in the absence of forward consideration, communities rebuild in a similar manner as before the disaster (Berke *et al.*, 1993).

Utilising data from two studies that examined relocation decision making in New York City following Hurricane Sandy, this chapter explores this decision-making process at both the household and community levels. Hurricane Sandy offered a unique opportunity to investigate this process due to the widespread destruction along the east coast of the US and the sizeable financial incentives the State of New York offered select residents to relocate away from coastal areas. Specifically, this chapter presents a case study of Oakwood Beach, a neighbourhood that served as a pilot site for a larger property acquisition programme (referred to as a buyout programme) managed by the State of New York.

We begin by outlining the history of Oakwood Beach and describing key characteristics of the community to provide context for the case study. Next, we recount the hazard history in Oakwood Beach, detailing both hazard impacts and previous mitigation efforts by the community. In the following sections, we summarise the effect Hurricane Sandy had on Oakwood Beach and discuss the subsequent reorganisation of the community and the buyout effort. We close by exploring both the implications of the buyout effort in Oakwood Beach on ensuing buyout efforts and future areas of research illuminated by this event.

Oakwood Beach

Oakwood Beach is a residential community located on the coast of Staten Island, one of New York City's five boroughs. Oakwood Beach was developed as farmland in the late 1600s (Lundrigan, 2004), but became popular as a beach vacation destination in the early twentieth century. Many of the homes in the area were originally built as beach cottages, which were converted to year-round residences during the Great Depression and after 1964 when the Verrazano-Narrows Bridge increased access to the island (Barr, 2013). For residents of New York, Oakwood Beach and the surrounding neighbourhoods offered the opportunity for people with modest incomes to own a detached home with a yard and a bit of privacy: a rare luxury in the city that contributed to a surge in development in the 1960s. While this development established Oakwood Beach as a permanent residential community, it also resulted in the destruction of many of the wetlands that offered natural mitigation against flooding events (Knafo and Shapiro, 2012).

Prior to Sandy, Oakwood Beach was a close-knit, working-class community, composed largely of long-time residents with deep ties to the area. Many residents described intergenerational ties to the community, and reported visiting Oakwood Beach as children, with their parents or grandparents, when it was still a summer beach community. Residents also described their community as unique, as a "paradise" in the city (Binder et al., 2015):

> We really did love where we lived. From the, uh . . . the wildlife, aside from the fish, turtles, frogs, aquatic life, there was this incredible amount of, um, water fowl, and other types of . . . of birds that bird watchers would actually come down and see.

While residents described their community in overwhelmingly positive terms, there was mention of challenges as well. Residents described negative connotations associated with their geographic setting and, by extension, their community, referred to as being "below the Boulevard", meaning Hylan Boulevard, the main road that designates the inland border of Oakwood (see Figure 14.1).

> You know, because when we bought the house, my husband always used to joke around, you know in Staten Island there's this saying "above the boulevard and below the boulevard." So, we were below the boulevard and he always used to joke around and say the Staten Islanders think that all, you know, it's like the wrong side of the tracks people live below the boulevard. I told him, "Oh boy, I guess we're just going to have to live with that stigma." [Laughs] You know, they're such great neighborhoods, but because they're known to have the flooding zones.

FIGURE 14.1 Map of Staten Island and Oakwood Beach. Oakwood Beach is a residential community surrounded by wetlands, and bordered by the ocean to the east and a wastewater treatment facility to the south. The community covers an area of just over 2.5 km², and sits an average of 1.5 metres above sea level. The worst of the flooding from Hurricane Sandy occurred in the area between the ocean and Hylan Boulevard, also affecting the neighbouring community of Oakwood. We outlined Census Tract 128.05 in black, and the area included in the buyout in grey. We created this map with the help of Hans Louis-Charles using ArcGIS® software by Esri. ArcGIS® and ArcMap™ are the intellectual property of Esri and are used herein under licence. Copyright © Esri. All rights reserved. For more information about Esri® software, please visit www.esri.com.

FIGURE 14.2 Typical Oakwood Beach home. Most homes in Oakwood Beach are converted beach bungalows, like the one pictured here. The converted bungalows are typically one- or two-storey, single-family, detached homes, with approximately six feet separating one house from the next. Newer construction in the community includes condominiums and a small number of newer, more traditional beachfront homes.

This phrase implied that outsiders viewed the area as somewhat less desirable, but residents repeated it without malice or shame. It did belie, however, an underlying awareness of the risk associated with living in a low-lying, waterfront area.[1] Incidentally, Hylan Boulevard was also the outer limit of major flooding from Sandy.

Hazard history

To understand Hurricane Sandy's impact on Oakwood Beach, it is important to understand the history of hazards for residents of the area. Minor but frequent hazards came with the territory. Oakwood Beach is low-lying, surrounded by wetlands, and occasional wildfires and small-scale localised flooding events were common in the area before Hurricane Sandy. As noted by one resident, the area exists in a perpetual state of "double trouble" (Binder *et al.*, 2015) when it comes to hazard exposure:

> We lived with constant threat of fires, secondary to people at the beaches. As well as the floods because at the end of our street was a creek. So if the flood

gates backed up then the creek backed up and then that came back into the street, so we were constantly at risk due to the fires and the floods.

While residents described these smaller-scale events as problematic, they were, for the most part, worth enduring for the benefits of residential living in the area. Wildfires, while a constant threat, were primarily small-scale events. The community's experience of coastal flooding events, however, was more complex. Along with minor flooding from routine storms, on several occasions the area faced the threat of flooding from hurricanes. Just a year before Sandy, experts warned residents that Hurricane Irene would bring extensive flooding to the area. In the end, however, Irene only caused minimal damage, and residents noted it as a non-event. A powerful nor'easter in December 1992 occupied the minds of residents as the most recent major flooding event to occur in Oakwood Beach. This storm brought three feet of storm surge and 130-km-per-hour winds through much of the area, damaging a majority of the homes and destroying a number of cars in the community (Paulsen, 2012).

Oakwood Beach Flood Victims Committee

The 1992 nor'easter brought, for many residents, the first instance of significant flooding to Oakwood Beach. Residents reported having from several inches to several feet of water in their homes and basements, which came as a shock to the community. After this storm, Oakwood Beach residents began to question, for the first time, the safety of their coastal community. In the wake of this storm, a group of residents formed a committee to discuss how the community should deal with these new concerns. The committee, which came to be called the Oakwood Beach Flood Victims Committee (FVC), began as an informal gathering of concerned neighbours. According to a former member of the committee, some residents expressed interest in selling their homes and relocating after the nor'easter, but in time the residents decided that pursuing a federally funded buyout was not the best course of action. Some residents were simply committed to staying in the area, while relatively low real-estate values at the time dissuaded others from selling.

Once they decided to rebuild, the FVC focused on lobbying for additional structural protections to mitigate against future flooding. A former resident of the FVC described the early phases of this effort:

> We started doing meetings. And, the first few meetings, everybody was really angry. Um, "We're not gonna put up with this anymore! We've been living like this for, you know, 50 years, and . . ." So, it was decided that we were going to petition the different levels of government, um, for repair and restoration to the shoreline, so that we could continue to live there.

The efforts of the FVC eventually achieved some success, though the process took nearly a decade. The most notable change involved repairs to a berm and seawall installed in the 1950s to protect the community from storm surge and, in 2000, the installation of a levee and tidal gate along Oakwood Creek, a manmade creek that separates the community from the ocean (ACE, 2013). Once these structures were repaired, residents reported feeling, once again, that their community was a reasonably safe place to live. With this, residents continued to invest in their homes and their lives in the community:

> My home took, from the bungalow, we, in 2001 decided to expand. Because of all these promises that were bein' made about the government about what was going to be done. To prevent this [flooding], or to downgrade it, at least.

Still, concerns remained that mitigation measures were never sufficient or properly maintained. The Army Corps of Engineers (ACE) had originally recommended replacing the existing seawall, but this was determined to be cost-prohibitive. They then issued a second set of recommendations, calling for a series of levees that would surround the community to guide future storm surges into a neighbouring wetland (Knafo and Shapiro, 2012; O'Grady, 1998). These plans, however, were never implemented. Instead, residents reported that a series of new homes were built on the land designated to become one of the levees:

> [U]nfortunately, because of funding, is what it was blamed on, um, we did not get the [flood] gate that was supposed to be operating, the way it was supposed to. We didn't get the height that was required for the berm . . . We didn't get the repairs to the seawall that their own study said that we needed. And we didn't get the second levee that their own study said that we needed to prevent flooding for this community.

Shortly after Sandy, the *Huffington Post* summarised Oakwood Beach's history as follows:

> For almost half a century, two generations of Staten Island's generally con-servative, independent and anti-government residents had petitioned the government to build and maintain a variety of barriers to defend them from the sea—and although elected officials had promised, on many occasions, to protect residents and their homes, authorities never delivered safety mea-sures that the government's own Army Corps of Engineers deemed worthy of serious consideration. The population simply continued to boom, and as real estate developers and their political allies pushed for growth, Oakwood Beach, like many other shoreline neighborhoods, morphed from a bungalow community into a modern suburb.
>
> *(Knafo and Shapiro, 2012)*

The Hurricane

Hurricane Sandy made landfall approximately 190 kilometres south of Oakwood Beach on 29 October 2012. The storm resulted in 24 deaths on Staten Island, including three in Oakwood Beach (Yates, 2013). Property damage in the community was extensive, with the storm damaging 909 structures in Oakwood Beach, resulting in more than 4 feet of flooding in one-quarter of those structures (New York City, 2014). Noted as one of the most heavily impacted neighbourhoods in New York, a number of media outlets called Oakwood Beach the "Ground Zero" of Hurricane Sandy damage (Knafo and Shapiro, 2012; New York Rising, 2013). Residents of Oakwood Beach encountered a 13–15-foot-high storm surge, and on Kissam Avenue near the coast, the storm surge ripped 13 of the 17 homes off their foundations. In addition to the direct damage from the storm, storm waters inundated the adjacent Oakwood Beach Wastewater Treatment Plant, resulting in the discharge of 237.5 million gallons of mostly treated sewage into the community (Kenward et al., 2013).

A number of residents noted that their experiences in the 1992 nor'easter and Hurricane Irene influenced their expectations for Hurricane Sandy. Irene

FIGURE 14.3 Former site of a home in Oakwood Beach. The home that previously occupied this empty lot was one of the first to be torn down after the buyout programme was implemented. The flowers in the middle of the lot were placed as a memorial to a father and son who drowned in this home during Hurricane Sandy.

was predicted to cause extensive damage in New York, prompting the City to issue a mandatory evacuation of the area for a storm that ultimately resulted in minimal damage in Oakwood Beach. Just over a year later, this experience served as a barometer for the expected impacts of Sandy. Residents noted that they did not take a number of the precautions for Sandy that they had taken prior to Irene (including evacuating, taping up windows, etc.) because of the lack of damage caused by Irene. With the apocalyptic portrayal of both events by the media and the City, residents found it difficult to imagine the level of damage Sandy inflicted as a possibility. As noted by one interviewee, "Based on our experience in 1992 and the non-event that was Irene, we spent the evening in a state of denial."

Organising, and the pursuit of a buyout

After returning to their community following Sandy, a small core of residents and property owners, including many of the original members of the FVC, met to discuss what they could do to help their community recover. They decided to repeat what they did after the 1992 nor'easter and organised a series of community meetings to explore existing needs and discuss recovery options. News spread by both word of mouth and flyers posted around Oakwood Beach, and about 200 residents gathered in a school auditorium in early November 2012. According to a number of community members, the group began considering the prospect of a buyout almost immediately. While they ultimately opted for structural protections after the 1992 nor'easter, many residents now discussed the buyout as a way out of an impossible situation, a chance for a fresh start, and an opportunity to avoid the constant flooding and fire risk that came with living in Oakwood Beach. The group began to call themselves the Oakwood Beach Buyout Committee, created a website, and started using prior data collected by ACE to formulate the case for a buyout for Oakwood Beach.

While many community members supported the prospect of a buyout, they also had significant concerns about the process. In general, concerns centred on three key issues. First, residents were worried that the amount the government would pay for their properties would not reflect the true value of their homes or be adequate to pay off their current mortgages and help them relocate. Second, although land purchased through buyouts is required to remain as open space, they suggested that the state would inevitably sell their land to developers in the near future, thereby profiting from their loss. Lastly, residents felt that their history in, and current ties to the area, coupled with the land and privacy they enjoyed in Oakwood Beach, were both invaluable and irreplaceable.

For a majority of the community members, anxiety about staying outweighed their concerns regarding the buyout programme. As details about the potential buyout programme materialised and residents learned they would receive pre-storm values for their homes, many concluded that if they did not participate in the buyout programme, they would not be offered better options in the future. They questioned the quality of life waiting for those who stayed in the community if a majority decided to accept a buyout offer. Specifically, residents cited concerns that crime would increase and city services would decrease in the area, or that the housing market would be permanently altered, leaving housing values so low that there would be little to no prospect of selling their homes in the future. Some residents were concerned that the government would forcibly take the property of residents who opted to stay:

> Like those people [who stay], they're not thinking what's gonna happen to services . . . Do you think the cops are gonna respond, Johnny on the spot? You know what I'm saying? And according to people in the neighbourhood, I never saw it, but the end of Tarlton they're supposedly having drug deals going on, and all this stuff going on down there. Nobody's living there, what you think's gonna happen? You think that's gonna *stop*?

Even if they wanted to rebuild, many residents were concerned that they would not have the financial resources necessary to do so. Although many residents carried homeowner and flood insurance policies, insurance payouts and aid packages were not sufficient to return their homes to pre-storm states. Beyond that, potential policy changes, including dramatically increased flood insurance premiums and expensive mitigation requirements, were looming and threatened to price them out of their own homes. Lastly, and perhaps most importantly, residents suggested that they no longer felt safe living in Oakwood Beach, and did not believe anything could be done to make the area safe again. While Sandy was the most devastating event in the area in recent history, it was also another event in a sequence of emergencies and disasters affecting the area. A portion of residents suggested that, even without a buyout, they would not return to Oakwood Beach:

> It wasn't just the storm. It wasn't just the storm that made people want to get bought out . . . I kind of likened it to a boxing match, where they were constantly getting beat up through every round . . . Now Irene comes, and that was what knocked everybody in this boxing match back on their heels. Because they were dealing with the floods, the natural floods from just the regular rain, the fires and everything like that, and now they have this ocean storm that brought both . . . But then when Sandy came, that was the knockout punch.

The experience of the buyout

On 25 February 2013, Governor Cuomo announced that Oakwood Beach would serve as a pilot community for a state-run buyout programme. The programme, federally funded through the US Department of Housing and Urban Development, offered homeowners pre-Sandy values for their homes, plus incentives to sell. While this announcement signalled a successful outcome for the community's efforts, significant difficulties remained. Residents described the time from the announcement of the buyout programme to the first property purchase as stressful and laden with uncertainty. A number of residents suggested that they were not convinced the buyout would actually occur until they received their cheques, and that they still had to find adequate interim housing, struggle for aid to sustain their households, and pay their mortgages until they completed the buyout process. As one resident stated, "Every step of the way it was a *fight*". This uncertainty continued for just over seven months, until the state purchased the first home in Oakwood Beach on 7 October 2013, just shy of the first anniversary of Sandy.[2]

In retrospect, many community members made positive appraisals of the buyout programme. They praised its simplicity, describing the buyout programme as an unambiguous, three-step process, as typified by the comment: "the NY state buyout was clear cut and expeditious". The story was, for the most part, consistent among residents. Community members suggested that they began by working with local community leaders to petition for the buyout. Next, they filled out an application with New York State for the buyout after the programme opened. Finally, they received a buyout offer from the state approximately 11–16 months after the storm.

Often, residents described the buyout not as one potential road to recovery, but as their only hope. "Thank God for a buyout program" was a common sentiment among participants. Community members suggested that, with the state of Oakwood Beach following Sandy and the seeming lack of a future they saw for the area, their reasons for staying in Oakwood Beach lost their salience: the destruction took with it any prospect of staying in the area.

> At that point, all of this stuff, raising their children and their grandchildren there, all of that stuff was meaningless because people died on the street and um, but um, you know, um, they realized that this whole thing that's happening is not going to go away, and all they wanted is they wanted, they wanted to just get out. They wanted to be made whole somehow. They were all thinking the same thing, that if there was any way for them to get out of this, they would get out of this. And when the buyout came, they all just . . . it was like a lifeboat for everybody and they wanted to get on.

Satisfaction with the process, however, was not universal. Some residents suggested that the amount paid by New York State was not adequate to find replacement-level housing on Staten Island, while others who lived just outside of the buyout zone questioned how the state developed inclusion criteria for the programme. A portion of residents suggested that the process was enigmatic, uncoordinated and stressful. They highlighted difficulties they encountered when trying to coordinate mortgage payments between ProSource Technologies, a consulting firm New York State commissioned to administer the buyout programme, and their personal banks:

> Very stressful was process with mortgage company: show that you have enough money on your account, they don't trust the contract of Buyout Programme "ReCreate NY" "ProSource" with written info which amount of money will be given to me. Eventually I am not happy because of all extra expenses which appeared because of moving out and in. My "dream" to pay off the mortgage of former house before retirement was gone and now my current house I will [be] unable to pay off. On the top of everything my marriage collapsed. It was too much for my husband.

Discussion

While the story of Oakwood Beach is, in many ways, unique, it provides a valuable case study of the process and impact of post-disaster relocation and home buyout programmes. Before Sandy, residents of Oakwood Beach had experienced a history riddled with natural disasters, and had organised and advocated for mitigation measures that would, they hoped, protect them from future storms. While these early efforts at mitigation ultimately proved unsuccessful, the community's social ties and early organising efforts were vital in the post-Sandy recovery. When Hurricane Sandy hit and the structural mitigation measures failed, the community drew on its disaster experience and organising history to advocate once again, this time for a home buyout programme that would permanently relocate them outside the area. While the community lacked a pre-Sandy formal recovery plan, their social ties and access to resources, established and fostered by previous community experiences and aided by their political acumen, served as a pseudo-recovery plan, enabling the community to act quickly and collectively when the need arose. The role of the community in the process is equally important to note. While there were outside forces at play in the implementation of the buyout in Oakwood Beach, this was, in large part, a community-initiated relocation. The decision by the majority of the residents of Oakwood Beach to relocate was perceived as a sensible—and in some cases, the only logical—next step in what can be understood as an extended recovery planning process. Had these conditions not been in place, it seems unlikely that the community would have arrived at the same outcome.

In considering what this may teach us about future buyouts, we see several community contextual factors that may contribute to the successful implementation of a buyout. Extant literature demonstrates the role of pre-disaster social conditions in shaping disaster recovery (Berke et al., 1993; Dynes, 1991; Fraser et al., 2003; Smith and Wenger, 2007), and our data indicate that post-disaster buyouts are no exception. In Oakwood Beach, strong social networks were associated with both the formation and work of the FVC after the 1992 nor'easter and the subsequent (re)formation of the Buyout Committee after Sandy. The role of these social ties was also evident in the community's collective pursuit of the buyout. While the decision to accept or reject the buyout was ultimately made independently by each household, with relatively few exceptions, the community moved to collectively accept the buyout. This suggests that community context had a strong influence on individual and household decisions. At the same time, participants were careful to point out that the decision to pursue a buyout was not an easy one. Although most residents eventually concluded that rebuilding in Oakwood Beach was not a viable option, this was a grievous decision that resulted in the loss of a home and a community that many loved.

As a final note, research on home buyout programmes is limited, and the research we presented here, in some ways, raises more questions than it answers. As the effects of climate change become apparent, more homes and communities will face the prospect of frequent or severe flooding. At a large scale, this raises the question of whether relocation is a difficult choice to be made by a few unlucky communities, or whether it is, in fact, an inevitable outcome for communities near coasts or on floodplains. If it is the latter, and the experience of Oakwood Beach marks the start of a trend, then it is imperative that researchers, policy makers and community members work together to understand with whom and in what contexts home buyouts are appropriate, and what it means to responsibly assist people in relocating out of hazard prone areas. While it is tempting to judge the success of buyout programmes according to how many households participate, for example, this assessment alone is far too narrow in its scope. For buyout participants, the decision to sell one's home is just one decision in a long and complex process that also includes determining where to move to, how to cope with the financial burden of relocating (beyond what the buyout programme covers), re-establishing social activities and networks, and possibly dealing with changes in employment and schools. Relocating may reduce residents' risk of experiencing natural hazards, but it is by no means clear whether their overall quality of life will improve, deteriorate or be largely unaffected by their relocation. More research is needed to explore the long-term impacts of these programmes on affected households and communities.

Notes

1 While the origins of this phrase are not completely clear, it may have stemmed from the Great New England Hurricane of 1938, a Category 3 hurricane that led to 700 fatalities in the New York and New England region (Mandia, 2013).
2 The state subsequently expanded the buyout to additional communities that suffered major losses following Sandy with a history of repeat flooding events, including Ocean Breeze and Graham Beach, Staten Island, and Suffolk County, Long Island.

References

ACE (2013) *Oakwood Beach, Staten Island, NY: Repair of Previously Constructed Projects.* Army Corps of Engineers. Available at: www.nan.usace.army.mil/Portals/37/docs/civil works/SandyFiles/Army Corps OakwoodBeach_FCCE_FactSheet.pdf.

Barr, Meghan (2013) 'Sandy buyouts in New York only affect few lucky homeowners in Staten Island', *Huffington Post*. Available at: www.huffingtonpost.com/2013/10/28/sandy-buyouts_n_4169483.html.

Berke, Philip, Kartez, Jack and Wenger, Dennis (1993) 'Recovery after disaster: Achieving sustainable development, mitigation and equity', *Disasters*, 17(2), pp. 93–109.

Binder, S.B., Baker, C.K. and Barile, J.P. (2015) 'Rebuild or relocate? Resilience and post-disaster decision-making after Hurricane Sandy', *American Journal of Community Psychology*, 56(1–2), pp. 180–196.

Dynes, Russell (1991) *Disaster Reduction: The Importance of Adequate Assumptions about Social Organization*. Disaster Research Center Preliminary Paper no. 172. Available at: http://udspace.udel.edu/bitstream/handle/19716/547/PP172.pdf;jsessionid=82AD8D28ED30C4C44592DC17EFAC3ACE?sequence=3.

Fraser, James, Elmore, Rebecca, Godschalk, David and Rohe, William (2003) *Implementing Floodplain Land Acquisition Programs in Urban Localities*. Chapel Hill, NC: The Center for Urban and Regional Studies, University of North Carolina at Chapel Hill FEMA.

Kenward, Alyson, Yawitz, Daniel and Raja, Urooj (2013) *Sewage Overflows from Hurricane Sandy*. Available at: www.climatecentral.org/pdfs/Sewage.pdf.

Knafo, Saki and Shapiro, Lila (2012) 'Staten Island's Hurricane Sandy damage sheds light on complicated political battle', *Huffington Post*. Available at: www.huffingtonpost.com/2012/12/06/staten-island-hurricane-sandy_n_2245523.html.

Lundrigan, Margaret (2004) *Staten Island: Isle of the Bay*. Mount Pleasant, SC: Arcadia Publishing.

Mandia, S.A. (2013) *The Long Island Express: The Great Hurricane of 1938*. Available at: www2.sunysuffolk.edu/mandias/38hurricane.

New York City (2014) 'The Staten Island Bluebelt: A natural solution to stormwater management'. Available at: www.nyc.gov/html/dep/html/dep_projects/bluebelt.shtml.

New York Rising (2013) *New York Rising Community Reconstruction Program: East and South Shore Staten Island Conceptual Plan*. Available at: http://stormrecovery.ny.gov/sites/default/files/crp/community/documents/staten_island_conceptual_plan.pdf.

O'Grady, J. (1998) 'Oakwood Beach: A levee is still a no-show', *New York Times* 1 November. Available at: www.nytimes.com/1998/11/01/nyregion/neighborhood-report-oakwood-beach-a-levee-is-still-a-no-show.html.

Oliver-Smith, Anthony (1996) 'Anthropological research on hazards and disasters', *Annual Review of Sociology*, 25, pp. 303–328.

Paulsen, Ken (2012) '20 years ago today: Historic nor'easter walloped Staten Island', Silive.com. Available at: www.silive.com/news/index.ssf/2012/12/20_years_ago_today_historic_no.html.

Perry, Ronald and Mushkatel, Alvin (1984) *Disaster Management: Warning Response and Community Relocation*. Westport, CT: Quorum Books.

Smith, Gavin and Wenger, Dennis (2007) 'Sustainable disaster recovery: Operationalizing an existing agenda'. In Rodriguez, H., Quarantelli, E. and Dynes, R. (eds) *Handbook of Disaster Research*. New York: Springer, pp. 234–257.

Yates, James (2013) 'Remembering 24 killed by Hurricane Sandy on Staten Island', Silive.com. Available at: www.silive.com/news/index.ssf/2013/10/remembering_hurricane_sandys_v.html.

15

CONCLUSIONS

Emerging lessons on community engagement in post-disaster recovery

Graham Marsh, Iftekhar Ahmed, Martin Mulligan,
Jenny Donovan and Steve Barton

In concluding this book, our aim is to highlight the fact that there is no 'one size fits all' approach to post-disaster/conflict community engagement in effective reconstruction and social recovery. Every disaster has different causes and consequences and every community is different. Every nation has differing methods of responding and the resources available to them will vary, as do the institutional structures. The levels of preparedness, funding and insurance available will also vary, with richer countries clearly being better able to cope financially with a disaster than poorer ones. It is often the poorest countries which suffer the most damage, as was the case following the Indian Ocean tsunami in 2004. The efficacy of humanitarian aid will face challenges in countries that do not have the resources to distribute aid effectively or corruption may interfere with how it is distributed. We decided to create this book because we know that everyone with a professional or scholarly interest in effective reconstruction and recovery can learn a lot from reflections on diverse practical experiences. However, we also understood that effective reconstruction and recovery work needs to take into account particular physical, social, cultural and political contexts within which the work is taking place. Climate change impacts and increasing global tensions are making disaster and conflict recovery work even more important, and the negative consequences of inappropriate or wasteful practices are multiplying. We conclude as we began, by stressing that good practice must start with detailed situation analysis and we can only hope to depict underlying principles rather than uniform models and approaches. We do not intend to say, therefore, that this or that method of rebuilding a shattered community is best. Each situation demands its own well-thought-out, informed and creative response. However, we have noticed several themes emerging from the various studies in this book that we hope the reader may find interesting, and with which those responsible for post-disaster or post-conflict resolution may find resonance.

Community engagement: nuances and dilemmas

All the practical experiences reported in the case study chapters confirm that every local 'community'—assuming such an entity actually exists—is different, having their own particular dynamics. In his chapter on Sri Lanka and India, for example, Mulligan concluded that: 'All local communities contain a host of sub-communities' and global flows of people, ideas and influences mean that they are more fluid than ever before. It is important to think of communities as processes rather than entities.'

Max Weber summed up the situation well when he referred to 'Communities of interest' rather than *the* community (Neuwirth, 1969). His argument was that in any locality there are a number of communities based on such elements as ethnicity, religion, class, caste, etc. which may have a great deal or a little in common with each other. There is the possibility of a largely homogeneous community in any particular locality, but this should not be taken for granted. Even in a single village, street or block of flats there may be a number of interested parties who may be in conflict with each other or who may have little or no contact with their neighbours. Furthermore, there may be deeply embedded political or language barriers which make the engagement of the community(ies) in post-disaster/conflict recovery processes difficult. The extreme example for where such engagement is problematic can be found in disaster sites where conflict is, or has been, the norm, and ethnic, religious or national differences undermine the prospects for community cohesion. The Rwanda, Philippines and Palestine chapters (Chapters 11, 10 and 9, respectively) are good examples of where this is the case, with tensions and divisions negating attempts to restore devastated areas effectively over the long term. MacLellan states: 'Fissures in Rwandan society, wrenched apart by the genocide, led to a breakdown in the structures of communities, both physical and psychological. The nature of the conflict and its consequences left a people traumatised by their experiences and mistrustful of others, resulting in a failure to fulfil the expectations of community, both traditional and instinctive.' By contrast, Donovan notes that warring parties in Palestine are 'resourceful and resilient communities who feel a deep connection to the land', but the different and conflicting connections of the two communities mean that what represents improvements for one community are detrimental to the other. In the Philippines, as Beza, Johnson and Fuentes point out, women stand out as being the bonding agents, bringing hope for stitching together conflict-torn communities.

At the other end of the spectrum, there are the nations where community engagement practices are more adaptable from one disaster site to another; for example, Australia and Canada, simply because the interests of the residents are not too divergent. Yet even in the Canadian example, there was substantial disagreement about the most effective way to rebuild physically, emotionally and economically. Even if not in actual conflict, residents in apparently cohesive local communities may be divided about how best to rebuild, how soon and where. While this book highlights many positive examples of where community engagement enhances the recovery processes, a number of the case study chapters indicate that too often it is the loudest voices, usually of the local elites, which are heard, while the needs of the more silent, perhaps the majority, go unheard or ignored. In Palestine, for example, the views of the women were traditionally under-represented, as it was the male leaders whose views were considered by the humani-tarian agencies. The Indonesian chapter (Chapter 7) provides an example of the consequences of the lack of consultation. There was no consultation with those people who survived the tsunami. Powerful players 'staked their turf' with the consequence that 'build back better' did not occur, and substandard housing was the result. To be fair, as O'Brien, Elliot and McNiven state, the sheer scale of the disaster made it difficult to mobilise communities.

Corruption in the political sphere also impacts upon the communities as they attempt to recover from disaster/conflict, as reflected in the Bangladesh chapter (Chapter 3). Corruption exists in many forms, as its scale and type vary from coun-try to country. While chapter authors have often anecdotally referred to corrupt practices, it was decided that this topic was best placed aside for later research, after more evidence had been gathered to substantiate the assertion that corruption is present. Its place in the recovery processes is, however, too often inescapable. For example, it can go from minor corruption (contracts awarded to help a family mem-ber out or because of social expectations rather than going to the best provider) to major corruption, with the syphoning-off of resources, often with the involvement of government and political elites. It is the editors' view that people working in disaster management and recovery should avoid getting caught up in corrupt prac-tices, but there is little that external people can do to address entrenched corruption in many of the countries that require disaster-recovery assistance. It is a matter of learning how to negotiate without engaging in corrupt practices.

One important observation from the chapters of this book is that while tension is often to be found between ethnic or religious groups, and indeed can cause or compound disasters, this is not always the case. This was demonstrated in the Indian case study (Chapter 6), where Muslims and Hindus lived harmoniously, assisting each other in the recovery processes. Similarly, in the Philippines participatory processes enabled Christian and Muslim women to collaborate successfully in a spirit of trust in their recovery processes and in moves towards a more peaceful society (Chapter 10). As Beza, Johnson and Fuentes note: 'The women are considered proficient commu-nicators and negotiators, working towards overcoming mistrust, as well as support-ing community-based recovery processes . . . Women's roles in this process are not promoted because of a lack of male influence, but rather because women are seen as contributing to the greater good of the village and the wider community.'

The case studies in this book suggest that when people enter disaster sites to undertake reconstruction and recovery work, they should challenge their own assumptions through prior research, and this applies to sites in countries like Can-ada as much as in countries such as Rwanda. The Canadian case study (Chapter 4) indicates that community engagement is often proclaimed without actually taking place; there is a one-way process of information dissemination, which does not meet real and diverse needs.

It is difficult for people who come from outside the disaster- or conflict-affected communities in which they find themselves to understand how these particular communities work and yet they are often obliged to act as conduits between the communities and the various layers of authority that enable relief, recovery and reconstruction work to proceed. Local knowledge is often a neglected asset, yet it can be complemented by various forms of largely external 'expert' knowledge. Western preferences for 'modern', 'scientific' knowledge may have little relevance in countries where other forms of knowledge are prevalent, and dispassionate knowledge may have little relevance when people are emotionally fragile. As Mulligan noted in his introductory chapter, 'it is much more difficult to work with complex, traumatised local communities than most practitioners imagine and very few of them have ever had any training in this work'. MacLellan in the Rwanda chapter (Chapter 11) highlights the difficulties for those working in such communities: 'The conflicting paradigms of ethnicity, behaviour and expectation, poverty, and psychological and societal trauma conflate, and have impeded reconstruction and rehabilitation. Experiences of the genocide and initiatives for the renaissance of a new Rwanda have contributed to a delayed restoration of "community".'

The case studies presented in this book suggest that post-disaster or conflict reconstruction and recovery work must be highly context-sensitive, and this applies to different contexts within a single country as well as between different countries. Bogdan, Bennett and Yumagulova (Chapter 4) mention that 'In Alberta, public involvement in natural resource management issues has tended to lag behind other provinces.' The Canadian example shows that even in nations with disaster management policies which pay lip service to the benefits of 'community engagement', practices commonly fall way short of the policy rhetoric. Processes that are called 'participatory' may in fact be top-down, one-way attempts at information dissemination, which do not even attempt to build on the existing or potential capacity of the affected communities. The opportunities relating to collective action in dealing with the aftermath of the disaster are then overlooked in favour of short-term gains.

Other examples of the top-down approach and its success or non-success in bringing about long-term recovery are to be found in the Italian and Chinese chapters (Chapters 8 and 5 respectively). In Italy, the communities were successfully engaged in one of the regions but not in the other. It would appear that the most successful long-term outcome was to be found where community engagement was in place. Lazzati notes that: 'people's involvement has helped to identify priorities and needs, formulating a vision for the long-term development of their own town, thus fostering social cohesion and sense of identity.' Whereas, even though the housing needs were addressed, 'the centralised top-down approach adopted in L'Aquila turned the community into a passive recipient, undermining people's self-reliance and resilience.'

The disaster in China (Chapter 5) raises issues that should concern all involved in disaster management, where the sheer scale of an event can have major implications for the recovery of a population. In China, the response of the governments at all levels could hardly have been anything but a top-down approach. The earthquake not only destroyed whole cities and their infrastructures but tens of thousands of people died. Trained officials and community leaders, who would normally be in charge of the recovery, were among the dead and essential physical resources were buried in the aftermath of the earthquake. Whole communities were decimated, making community engagement in the immediate recovery phase almost impossible. So immense was the recovery task that for the first time, apparently, the role of NGOs was recognised by the Chinese government and various organisations were involved during all phases of the rebuilding and relocation. The model used there could also be adopted in other situations. Cities not impacted were paired with those which had suffered—with positive outcomes. However, not all of the outcomes were successful: many of the people were unhappy with what was built in the resettled areas, and the limitations of the community consultation were evident in the aftermath of the disaster and in the ongoing recovery processes.

While participation varies from one context to another, 'local communities' are never 'fixed entities' because of many interacting 'components', and are constantly changing in relation to broader social, economic and cultural contexts. It is important, then, to work with a dynamic understanding of 'community formation'. Communities operate at scales, ranging from the local to the national, and this can present challenges when disasters affect a range of local communities at the same time. In relation to flooding in Bangladesh, for example, Ahmed notes (Chapter 3) that a single event impacted '35 million people—a quarter of Bangladesh's population—and 4 million houses were damaged or destroyed'. And in India, as noted by Vahanvati (Chapter 6): 'The earthquake caused nearly 20,000 deaths . . . and destroyed over 1 million houses.' In the China study (Chapter 5), Chen points out that the impact was even more catastrophic: 'the earthquake resulted in 69,266 deaths, 374,643 injured and 17,923 missing . . . More than 15 million people were affected and had to be evacuated from their original homes . . . The total financial loss from the earthquake was 845.1 billion RMB.' The 2004 Indian Ocean tsunami killed more than 170,000 Indonesians on the island of Sumatra alone and left another 500,000 homeless. While we conclude that a 'one-size-fits-all' approach ignores a host of contextual considerations, relief and recovery agencies often have to work across numerous local communities affected by a single disaster and this poses the need to quickly replicate practices which appear to meet community needs and aspirations. Being context-sensitive does not rule out the potential to replicate good practices.

The complexities associated with providing effective support to disaster- or conflict-affected communities multiply when we consider the number of agencies that are likely to be involved in the processes and the diverse assumptions and practices that these agencies might employ. Following the tsunami in Indonesia, more than 120 international aid agencies joined 430 Indonesian NGOs to offer immediate assistance. In his Bangladesh case study (Chapter 3), for example, Ahmed notes that 'In its basic form, it was a tripartite partnership between UNDP, PNGOs and communities. However, the stakeholder network was much wider, also involving the national government and local government authorities at different administrative tiers, particularly in local level needs assessments, consultations, beneficiary selection, etc.' Ahmed goes on to conclude that there is a major difference in the dilemmas faced by the outside agencies when the scale of the event is so vast as against what is involved in dealing with more localised events. He suggests that 'community engagement has to be understood differently from a small-scale grassroots community-based programme. For direct community engagement in a large programme, more staff and resources are required and it becomes impractical for any one agency, even for a significant agency such as UNDP, to operate alone. This creates the need for the partnership strategy involving multiple agencies, and community engagement becomes more feasible when the partners have existing links and experience of working with the communities concerned . . . However, this strategy requires intensive monitoring, technical supervision, quality control and, importantly in a context such as Bangladesh, dealing with corruption.'

The Palestine case study (Chapter 9) shows that NGOs often compete with each other for funding, with client communities at the mercy of the outcome of that competition. Too often it is the case that scarce resources are distributed to politically favoured citizens, while others in greater need receive fewer or no resources, or have to wait longer to receive them. Many of the case studies in this book note that various forms of corruption divert resources away from disaster- or conflict-affected communities, but this may not be something that can be addressed by relief and recovery agencies, especially those who come from outside the nation concerned. Nevertheless, unless the processes are managed carefully, competing interests can impede the recovery and in some cases stall it completely. All involved, at all levels, should be encouraged to work in partnerships with one another, as occurred successfully in many of the countries. Important for this process is that community leaders need to be identified even if they do not occupy formal leadership positions, and external agencies need to work with a range of community leaders and representatives. Perhaps the overriding principle here is that there needs to be a spirit of cooperation with associated transparency, trust and a lack of bias built into the community engagement processes.

The Canadian case study (Chapter 4) highlights the fact that timing is all important when it comes to the processes of community engagement. Immediately following a disaster, traumatised survivors are not likely to be ready to discuss long-term reconstruction and recovery options. As noted in relation to post-tsunami recovery in Sri Lanka, it is critical to distinguish the community engagement needs of short-term relief from longer-term reconstruction and recovery.

No doubt the most difficult dilemmas are faced by those involved in recovery work when natural disasters are compounded by human conflict impacts. In such situations, the stresses placed on people are multiplied and compounded and they are psychological and emotional as much as physical or economic. MacLellan in the Rwanda case study (Chapter 11) examines the most challenging circumstances of all, when local communities have been torn asunder by cascading practices of genocide: 'The value of strong social networks, the "social capital" of individuals and communities, must be acknowledged and encouraged. Building such capital as a robust foundation for community relations would enhance reconciliation, reconstruction and reintegration in cases of conflict or natural disasters and is a significant part of peace building. Social capital should be secure and resilient before a crisis, so that it becomes more difficult to break down during and after an event.'

Physical rebuilding techniques: What worked and what didn't?

What is evident from these chapters is that where the community is effectively engaged in rebuilding the end result is more positive, particularly when locally appropriate options are possible. This often involved the use of local knowledge, materials and labour, where available, coupled with expert knowledge from the outside. A good example of how participation can lead to a successful outcome is to be found in the chapter on India (Chapter 6), where increased 'trust' between all participants involved was also an essential element in disaster recovery. The outside expertise was coupled with local knowledge and collaboration both during and after the rebuilding processes. Vahanvati concludes: 'Three lessons have emerged from the examination of four good-practice ODR [Owner Driven Reconstruction] case-study projects in India:

1) Gaining community trust and local partnership —a foundation for ODR.
2) Artisanal skills training or capacity building—during ODR.
3) Continued support for enhanced community self-sufficient—post-ODR.'

The Solomon Islands (Chapter 12) also provide a useful example of the role of participation in achieving successful outcomes: 'Communities are usually deeply destabilised by disaster, with day-to-day confidence undermined. The inclination of agencies to assist and advise can deepen a sense of helplessness at the exact moment when confidence in their own capacity needs to be restored.' Barton, the author, in order to deal with this dilemma, has developed what he terms a 'Beneficiary Driven Recovery [which] seeks to address these issues, first by making it clear that there are varying degrees of loss or impact and by suggesting a method to quantify this. The second step is to make the process of resource allocation transparent and equitable. Having established this, the process places selection of support squarely in the hands of the communities and their members.'

The opposite occurred in Indonesia (Chapter 7), where the recovery processes were actually hindered through a lack of consultation and an absence of cultural understanding within the aid agency. The recovery task in Aceh was certainly immense, but in this case it was not that a 'western' aid agency imposed its methods on the communities, but rather one from another 'eastern' country. This agency apparently failed to consult before building housing for the 'generic resident', allocated by a lottery as to where the residents would live, with the materials used including asbestos. The residents, totally unaware of any dangers, were not warned of the health risks associated with such materials. All these factors made recovery difficult for a traumatised people. As with many cases across the globe, the emphasis associated with rebuilding centred on quantity rather than the quality of the housing.

No matter what the context, a key aim should usually be to 'build back better (safer)', with not only the structures being more disaster-resistant and appropriate for the site, where possible, but with the well-being and future safety of the residents involved considered throughout the process. All residents, not just the male leaders, as was the case in Palestine (Chapter 9), should be consulted, educated and supported throughout all phases of the rebuilding. As discussed in the Australian chapter by Ireton and Ahmed (Chapter 2), concerns of the affected residents about such matters as 'the nature of their community into the future and the opportunities presented in rebuilding that may incorporate improved safety and environmental, community, aesthetic or economic outcomes', should be taken into account by relevant authorities and NGOs.

But programming around the slogan 'build back better' may not always be the most appropriate option. In the first place, we need to consider whether the residents in fact want to rebuild their houses or stay at the same location. This is particularly the case where there is growing concern for the future of their locality over the impending impacts on it of climate change. As Ireton and Ahmed, the Australian authors, indicate, 'A pre-conceived idea that often prevails within government and the media is that anyone who has lost a house to a bushfire (or other disaster) will naturally want to rebuild. There seems to be little thought about whether these people have prior experience in building a house, or have ever had the need or desire to build a house.' And this is usually the case even without concerns over what the impacts will be for them as the planet warms up in the near future. The psychological impacts associated with the disaster need to be taken into account as the pressures to rebuild can also be immense and traumatic for all concerned. Particularly so when the destroyed house, for example, may be in a disaster-prone area. In the Hurricane Sandy case in the United States (Chapter 14), where future impacts of climate change, with the potential for an increased number of hurricanes with severe surges and flooding, were in people's minds, prior experience of disaster made it easier for the residents to decide that the most sensible option was *not* to rebuild but to relocate: 'When Hurricane Sandy hit and the structural mitigation measures failed, the community drew on its disaster experience and organising history to advocate once again, this time for a home buyout programme that would permanently relocate them outside the area'. Yet, as Greer and Binder note: 'had these conditions not been in place, it seems unlikely that the community would have arrived at the same outcome.' They also state that: 'Relocating properties, while a seemingly logical solution, is not easy. Not only is it costly on the front end, the window for meaningful change is narrow. Support for such measures is often strongest immediately after an event, and in the absence of forward consideration, communities rebuild in a similar manner as before the disaster.' However, as they also note, there is no guarantee as to 'whether their overall quality of life will improve, deteriorate, or be largely unaffected by their relocation.'

Importantly then, the needs, the desires and the psychological state of the residents must be considered as integral parts of the processes of rebuilding and recovery. Referring to the Canadian research again: 'while community authorities may have been aware of the psychological phases of disaster response, in practicality, the findings suggest that they did not seem to incorporate this awareness into the timing or their approaches to community consultation.' In the future, Bogdan, Bennett and Yumagulova note: 'a variety of approaches that are supportive and reflective of the emotional phases of recovery would be advantageous as part of the pre-disaster recovery planning process.' As some of the chapters have highlighted, resilience and recovery should be about institutions that are accountable to serve disaster-affected communities and to maximise the potential for their engagement in all aspects of the recovery processes. There is a need, then, to ensure that the needs of the residents/community are not neglected, a point recognised by Vahan-vati: 'Three key lessons were discussed and proposed in the form of a new operational framework to operationalise community participation. The first finding was the significance of social process, such as grassroots motivation of survivors during the initial planning phase, even before the beginning of construction. The second finding, about building the capacity of the local community, is nothing new for the practitioner in disaster recovery, yet it is often compromised. The third finding was about planning beyond one project life cycle—that is, beyond rebuilding of houses—so as to ensure self-reliance of community in terms of livelihood, awareness and safe construction skills.'

Key lessons learned

One lesson learned from these chapters is that intervention by the authorities is not always received positively by the affected residents. In the Solomon Islands (Chapter 12), for example, the author concludes: 'On more than one occasion, community members suggested that they might be in a better state seven months after the disaster if aid agencies had never appeared. The provision of support to selected households under a system that was of uncertain "fairness" and not at all clear to the community members had disrupted the delicate balance of shared assets and responsibilities. Transparency, equity and fairness (as perceived by the affected population) are critical aspects of an effective response.' A further point that Barton makes is regarding how soon the rebuilding should take place. If done in haste—in order, for example, to rehouse the affected residents—the end result may be poor-quality, inadequate housing, or houses that will not be lived in once built. This author's 'extensive experience suggests that recovery support is usually a longer-term process and the risk of delay is outweighed by the risk of low-value or even inappropriate assistance.' This is also a key point in the Sri Lanka chapter, as pointed out succinctly by Mulligan: 'haste was the biggest detriment for long-term social recovery' (Mulligan, 2013, p. 280).

Consequently, it would appear to be essential for the authorities and NGOs to work constructively with the affected residents not only during the short-term recovery period but over the long haul as well. There is a need to recognise precisely what disaster recovery is, and the short- and long-term consequences of what is being done for and with those who have been impacted. In the Sri Lanka and India chapter, Mulligan concludes: 'many humanitarian agencies failed to appreciate the difference between relief and recovery, and many withdrew without putting any kind of "transitional arrangements" in place.' Later, he adds: 'Local knowledge is critical, and it is important that external aid agencies find ways to work with local people and community-based organisations in regard to resettlement planning. Effective partnerships can be forged between external agencies and community-based organisations, and all aid agencies should make sure that they have put in place adequate transitional arrangements before they withdraw from the communities in which they have worked. In short, the study recommends a more deliberative approach in moving from short-term relief to long-term recovery.'

Although a diverse array of cases studies, the key lessons emerging from the book's chapters point to a set of guiding principles for good policies and practice for the future, all of which hinge on the need to be context-sensitive. These principles for effective post-disaster community engagement include the following:

- Remembering that people with the best intentions can sometimes trigger or amplify tension or conflict within traumatised communities.
- Making as few assumptions as possible. Everywhere is unique. Being informed— through reading or consultation—about the history, culture, structures of authority and dynamics of communities being worked with is important.
- Confirming that understanding with the people with and for whom the work will be undertaken.
- Remembering that all local communities include sub-communities or interest groups which may have external linkages, so there is a need to proceed cautiously and consult widely.
- Recognising the landscape of influence, and identifying people who can represent groups of people within the community or communities, even if they do not hold formal positions.
- Looking for people who have particular skills or abilities which can create better options for those people and/or their community at large.
- Proceeding patiently, in case the work causes unforeseen resentment or resistance.
- Ensuring that the criteria being used to address particular needs are always made known to the community.
- Being aware that people, families and groups may face difficult choices about their futures after a disaster, and therefore avoiding the cultivation of expectations that are unrealistic or may limit their options.
- Looking for creative and culturally appropriate ways to help build an inclusive sense of community for those who have been traumatised by a disaster.
- Remembering that community consultations and engagement are processes over time rather than discrete events. Outside participants are likely to see the process of recovery for only a relatively short time. They may give it a momentum that might make it easier or harder for the community(ies) to progress down the road to recovery.
- Outsiders can also have the ability to change expectations and cultivate hope. Without this, locally generated recovery becomes much harder.

To conclude: there is a lot to be learned by examining practices in a wide range of geographic, social and cultural settings to reveal what has proved effective and what hasn't at helping people overcome the impacts of disasters. However, perhaps more important a message is not so much about techniques that may prove useful, as about the underlying approach needed to assist recovery and renewal. All communities are unique, influenced by geography, social processes, the momentum of history and the unique perspectives, experiences and capacities of the people who make them up. By extension, all places are experienced by people who hold a diversity of views and values and are tied to each other and their physical surroundings in many different ways, making the notion of a single community in a single place unreliable. Consequently, whatever techniques are used, the practitioner should not assume once a community has been identified that this is the *only* community. Care needs to be taken to search out and respect the perspectives behind those that are put most loudly.

Furthermore, disasters are by their nature extraordinary events, disrupting the familiar and often challenging the landscape of authority in the places where they occur. They destroy the social and physical fabric of the people impacted by them. If outside experts are to help these people to go down their own paths of recovery, then these experts should remember that what worked elsewhere may not work here. To paraphrase the Australian academic Trevor Budge: 'Once you have seen one disaster, you have seen one disaster.' The practices outlined in this book reveal that there are some underlying principles for good practice which are context-sensitive, but there are no universal models. At the end of the day, there is no substitute for humility, patience and an ability to listen to a wide range of voices.

If this does happen and the outside expert can respectfully recognise the diverse range of relationships between people and place, and the diverse ways these have been affected by the disaster, then they will be in a better position to tailor their input to what is needed. Only then, with relevance and respect, can the outside expert become a catalyst for hope and create a sense that a future worth striving for is possible.

References

Mulligan, M. (2013) 'Rebuilding communities after disasters: Lessons from the tsunami disaster in Sri Lanka', *Global Policy*, 4(3) (September), pp. 278–287.

Neuwirth, G., (1969) 'A Weberian outline of a theory of community: Its application to the "Dark Ghetto"', *British Journal of Sociology*, 20(2), pp. 148–163.

INDEX